Deep Learning for Deblending Crowded Star Fields: The Future of Astronomy

Sam K

Deep Learning for Deblending Crowded Star Fields:
The Future of Astronomy
Copyright © 2023 by Sam K

The first edition was published in 2023
ISBN:
Published by:
Star
1663 Liberty Drive
Hyderabad, IN 47403
www.starpublishers.com

This book is self-published using on-demand printing and publishing, which allows it to be printed and distributed globally.

TABLE OF CONTENTS

1
INTRODUCTION

In 2020, the Nobel Prize in Physics was jointly attributed to three physicists. Two of them, Reinhard Genzel and Andrea Ghez, for finding an extremely heavy and invisible object that pulls on a group of stars, causing them to orbit it at extreme speeds, the black hole at the centre of our galaxy: Sagittarius A*. In the press release, the chair of the Nobel Committee for Physics, David Haviland, further adds that "(…) these exotic objects still pose many questions that beg for answers and motivate future research. Not only questions about their inner structure, but also about how to test our theory of gravity under the extreme conditions in the immediate vicinity of a black hole." [1].

1.1. CONTEXT

The point spread function (PSF) represents the response of the acquisition instrument to a point source (e.g., a star at a very large distance). The approximation of stellar signals by a model PSF has proven to be extremely successful over recent decades and enabled multiple discoveries and breakthroughs in astronomy. However, and even though these methods have been extremely successful since their development, as better image acquisition equipment is developed and technological advancements are made, their limitations become more apparent. They are particularly noticeable when these methods applied to crowded star fields or regions of space where most objects are so dim that their brightness is difficult to distinguish from sky background noise.

Figure 1 illustrates an example of such crowded regions of space, where the signals of neighbouring stars may overlap. This makes their detection by conventional methods, such as PSF fitting, difficult [2]. In recent years, computer vision has witnessed significant advancements in object detection and instance segmentation with the rapid development of deep convolutional neural networks [3]. These techniques have seen broad adoption in astronomy, where one of the most prominent uses has been their application to tasks of deblending crowded star fields [4].

The goal of this study is to ascertain whether novel techniques, namely deep learning methods developed for object detection and instance segmentation, can achieve better results than the more widely adopted conventional techniques. Although these new methods are typically developed for the detection of generic or everyday objects, they have seen increasing success when applied to the most diverse areas such as healthcare, automotive, and agriculture, amongst many others [5]. Techniques such as transfer learning been one of the key factors for this success, allowing for increasingly complex models trained on very large and generic datasets by companies and institutions with vast computational power and resources, which small organizations or individuals can later adapt to more specific problems [6].

Figure 1 – Field of stars -captured with Wide Field Camera of Hubble's Advanced Camera for Surveys [7].

1.2. MOTIVATION

Black holes are one of the few unique testing grounds where core ideas of theoretical physics can be verified. Their immediate surroundings, are one of the few places in the universe where the unique properties of strong field gravity can be observed [8]. The ability to detect more and fainter objects in these regions of space would allow scientists to more accurately characterize regions in the vicinity of black holes, using this new information to either validate existing theories that explain how our universe works (such as Einstein Theory of Relativity) or, if any end up being proven wrong, to serve as a steppingstone from which to build new models that better explain how our universe works.

Since black holes do not emit any light, one of the few ways for scientists to study them is via their interactions with orbiting stars. Stars serve as probes into the behaviour of these massive objects and have allowed physicists to indirectly estimate the mass distribution in the Galactic Centre [9]. Precise imaging of this region is of utmost importance to propel scientific advancements. One of the critical tasks is the discovery of faint, unknown stars on short orbits that could potentially restrict the black hole spin [10].

1.3. OBJECTIVES AND PROJECT STRUCTURE

The key objective of this study is to understand whether the application of deep learning to the problem of stellar detection and photometric analysis can yield better results than the traditional techniques based on analytical methods. To achieve it, the following tasks are of critical importance for the astronomical problem and will form the backbone of the present study:

- Detection of stars, and in particular, faint stars that are difficult to distinguish from background noise and other perturbations.

- Accurate deblending of stars with small angular separation (i.e., overlapping point sources).

- The correct characterization of the position of the centre of intensity for each of the detected stars.

- Accurate measurement of the flux emitted by each star.

The literature review and state of the art study (Chapters 2 and 3) aim at: providing a summary of the key techniques that have been and are still widely adopted for the modelling and detection of stellar objects; performing an examination of novel deep learning techniques that have seen increasing success in detection and segmentation tasks and could be adapted to the task of stellar detection; and finally study cases where these artificial intelligence techniques have been used to solve similar astronomical problems of detecting stars, galaxies and other cosmological objects.

Once the relevant literature has been reviewed, the methodology (Chapter 4) of the study will be described, highlighting the key assumptions taken during development, and any important remarks regarding the direction chosen for the project. Subsequently, the result phase (Chapter 5) will summarize the main remarks found, providing an extensive comparison between the novel and traditional methods. In contrast, the discussion (Chapter 6) will highlight some of the key discoveries made during the study and provide further context and explanation where appropriate. Finally, the conclusion (Chapter 7) will summarize the main findings of the study, identify any shortcomings and provide ideas for future projects.

2
BACKGROUND

2
BACKGROUND

The Airy disk represents the profile of the diffraction pattern, or PSF, of a point source of radiation when imaged through an acquisition instrument, such as a telescope. George Airy proposed it in 1834, establishing that *"the image of a star will not be a point but a circle surrounded by a series of bright rings. The angular diameter of these will depend on nothing but the aperture of the telescope"*. This became one of the key concepts for fields such as astronomy and microscopy, being still widely used today [11, 12]. Given a telescope with an aperture a, imaging at a wavelength λ, the intensity I at an angle θ from the optical axis, relative to the intensity I_0 at the centre, is given by equation(2.1):

$$\frac{I(x)}{I_0} = \left(\frac{2J_1(x)}{x}\right)^2 \tag{2.1}$$

where $x = (2\pi/\lambda)\, a \, \sin\theta$ and J_1 is the first order Bessel function of the first kind (solutions to the Bessel differential equation). Rings can be noted at the maxima $x = x_1, x_2, \dots$ of $I(x)$. Considering the limit of small angles, where $\sin\theta \approx \theta$, these maxima would correspond to $\theta = \lambda x_i/(2\pi a)$, therefore the angular diameters of the rings are inversely proportional to the aperture a [12], as can be seen in Figure 2.

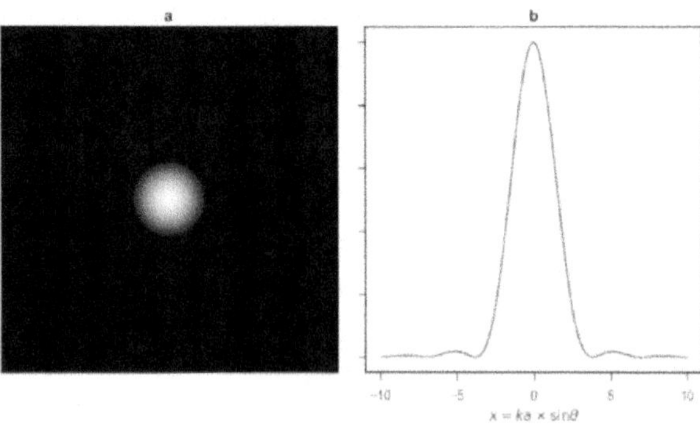

Figure 2 – Point source generating an Airy spot (a) with the corresponding intensity profile shown in (b) and derived via equation (2.1), where $k = 2\pi/\lambda$ [12].

When the angular separation between two point-sources decreases below a certain threshold, the lens system will form an image consisting of a distribution of partially overlapping, yet independent, Airy patterns. Considering the case of two stars seen through the objective lens of a telescope, their diffraction spots would be considered resolved if their centres were further apart than the centre of a single spot from its first zero intensity[1] [12, 13]. Figure 3 illustrates this phenomenon for different degrees of separation between point sources.

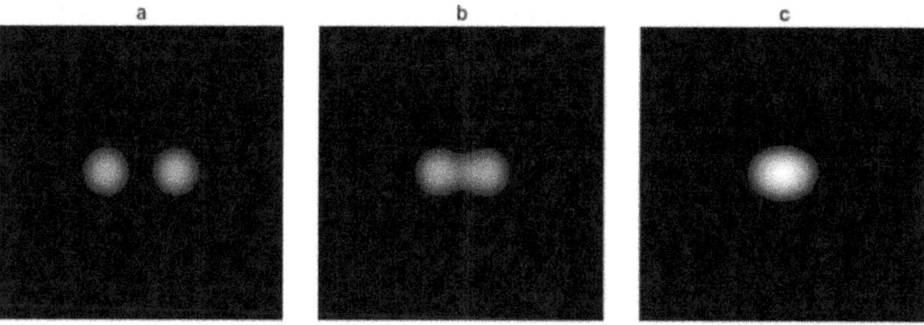

Figure 3 – Airy diffraction spots of fully resolved (a), barely resolved (b) and unresolved (c) point sources [12].

The problem of separating neighbouring sources, particularly in crowded star fields, is central to the field of stellar photometry and, alongside that of the aforementioned faint sources, is one of the critical tasks that many of the developed photometric methods attempt to solve. In observations from ground-based telescopes the PSF is further degraded by the turbulent atmosphere. The net effect is the blurring of Airy rings and a final function that is approximately Gaussian, but not exactly.

2.1. STELLAR OBJECT DETECTION WITH CLASSICAL TECHNIQUES

In a review of source detection approaches in astronomical images [2], the authors highlight the challenge of separating pixels that belong to astronomical objects from those that are simply due to the background or noise as one of the key obstacles. Despite the study focusing on the detection of all cosmological bodies (i.e., stars, clusters, galaxies, etc.) the principles are still applicable to the branch of stellar photometry. A three-step process is described: the first step consists of the detection of an astronomical object (also involving determining its position on the image), secondly, the classification of the object is performed, and finally photometric techniques are employed to determine the properties of interest, such as flux, magnitude, or intensity. A study on probabilistic catalogues for crowded stellar fields [14], highlights some of the standard tools of source detection that have been more widely used by astronomers over the recent decades, with the first being DAOPHOT.

2.1.1. DAOPHOT

Proposed in the late 1980s [15], DAOPHOT has become one of the more widely adopted methods in stellar photometry[2]. In his publication, the author states the problem that the algorithm aims to solve as: *"many interesting populations of stars for which photometric measurements would be useful are so densely packed on the sky that any given star is likely to be blended with its neighbours"*. Therefore, the proposed solution is to fit a *"superposition of predicted stellar distributions, plus a model for the diffuse night-sky luminosity, to a two-dimensional digital image obtained by a photometrically linear detector"*.

---

[1] The first zero intensity or extent of the Airy disk is the distance from the central, high intensity, region to the first trough that surrounds it.

[2] At the time of writing, the original paper has collected over 4300 citations, with 1000 of them being made after 2015, pointing to its widespread use in the astronomical community, more than three decades after its original proposal.

Diving more in-depth into the way that DAOPHOT operates, the author divides it into four main tasks: the first is to find a list of probable star locations; the second performs estimates of sky background; the third derives a model of the PSF for the frame; finally, photometry is obtained for all stars in the image via least-squares profile fits.

The process of detecting stars begins with an iteration through every pixel in the image. At each position, the program fits a two-dimensional Gaussian approximation to the PSF that the star would generate. If a star happens to be centred on that pixel, the Gaussian approximation will be good, and the central height of the fitted curve will be proportional to the brightness of the star. In contrast, if the pixel lies in an empty region of the sky or in the wings of another star, then the central heigh of the best fitting Gaussian will be close to 0, or even negative. In this way, the probability that a star is centred on a given pixel, is related to the relative height of the best-fitting Gaussian profile [15].

The bivariate circular Gaussian function G proposed is detailed in equation (2.2):

$$G(\Delta i, \Delta j, \sigma) = e^{-(\Delta i^2 + \Delta j^2)/2\sigma^2} \tag{2.2}$$

where Δi and Δj would reflect the difference between a neighbouring pixel i, j and the centre pixel i_0, j_0 and σ the standard deviation of the Gaussian distribution. Being the central brightness of the stellar profile that best fits the pixels around point i_0, j_0 given by H_{i_0,j_0}, and the estimate of the local background brightness given by b, then the observed brightness of a given pixel $D_{i,j}$ is given by equation (2.3):

$$D_{i,j} \doteq H_{i_0,j_0} G(i - i_0, j - j_0, \sigma) + b, \qquad (i,j) \; near \; (i_0, j_0) \tag{2.3}$$

where the symbol "$\doteq$" indicates a least squares fit to the data. The numerical value of H_{i_0,j_0} is finally given by linear least squares detailed in equation (2.4), where n is the number of pixels used for fitting:

$$H_{i_0,j_0} = \frac{\sum(GD) - (\sum G)(\sum D)/n}{\sum(G^2) - (\sum G)^2/n} \tag{2.4}$$

The second task focuses on estimating the sky brightness. The amount of light received exclusive of a particular star is independently estimated from a nearby region and the difference between the two (the amount of light found in the star region minus the amount of light found in a background region) provides an estimate of the amount of light received by the star alone. DAOPHOT estimates this by considering a region of the frame that is near a hypothetical star but far enough away so that only a negligible fraction of the stars' light falls within it. To sample the value of the background sky, a circular annulus centred at the star location is considered, with an inner radius several times the full width at half maximum (FWHM), and the outer radius at a distance such that between the two radiuses, the total number of pixels is at least three times the number of pixels of the aperture. The mode of the pixel values in that region is taken as an estimate of the sky brightness per pixel for that part of the frame.

The third part, the PSF routine, starts by fitting an analytic bivariate Gaussian function to the central region of a bright star. For each pixel, the two-dimensional integral of the Gaussian function over the area of the pixel is computed, and both the centroid and height of the Gaussian function are determined from the actual intensity values as detailed in equation (2.5):

$$D_{i,j} - sky \doteq H \int_{i-\frac{1}{2}}^{i+\frac{1}{2}} \int_{j-\frac{1}{2}}^{j+\frac{1}{2}} G(x - x_0, y - y_0, \sigma_x \sigma_y)\, dx, dy, \qquad (i,j) \; near \; (i_0, j_0) \tag{2.5}$$

which is solved for H, x_0, y_0, σ_x and σ_y, serving as a first-order approximation to the actual stellar profile. The observed residuals (difference between the best fitting profile and actual shape) are computed at each pixel location and used to perform empirical corrections for the function approximation to the observed profile.

The fourth and final step is defined as profile-fitting photometry. The program starts by defining a critical distance between two stars under which they should be reduced together, defined as the distance between the centroid of the brightest star and a point where its profile is indistinguishable from background noise. From there, the program iterates over each star and estimates the best fitting PSF. This estimate is subtracted from the original image and its residuals used for incremental corrections to the PSF profile. After reaching the last star in the image, the program obtains an array of residuals containing the sky brightness, random noise, and systematic errors due to inaccuracies in the estimates of the stellar parameters. This can be used to correct the stellar parameters and to create new estimates for the next iteration.

The author concludes the publication by identifying some of the shortcomings and potential improvements of DAOPHOT. One of the key issues identified is the difficulty in evaluating the diffuse component of the observed brightness at a given position in a crowded star field, highlighting that even the term *"diffuse sky brightness cannot be perfectly defined because different things are meant under different circumstances"*. Further limitations observed include the imprecisions to the PSF caused by optical aberrations or charge-transfer imperfections in the detector, the difficulties caused by particularly crowded star fields where the signals of multiple stars overlap, and the incapability for cross-identifying stars from one frame to another.

2.1.2. SEXTRACTOR

Another popular source extraction software is SExtractor[3]. Although overlapping some functionality with DAOPHOT, its focus is not solely on stars, but is instead capable of detecting a range of atronomical objects such as stars, clusters, galaxies, etc. [14]. Proposed by [16] SExtractor analyses a complete image in six steps: estimation of the sky background, detection (by thresholding), deblending, filtering of the detections, photometry, and star/galaxy separation.

In the first step, background estimation, the program starts by generating a "background map". To construct it, images are divided into a grid that is typically 32 to 128 pixels wide, and an estimation of the background values at each division is computed. An algorithm similar to the one used in DAOPHOT (see Chapter 2.1.1) is used, where the local background histogram is clipped iteratively until convergence at a $\pm 3\sigma$ from its median is achieved. Possible over-estimations due to background noise caused by extremely bright stars are addressed by applying a median filter to the grid. The resulting background map is a bilinear interpolation between the meshes of the grid.

For the detection step, a second pass through the image is performed. The authors selected the thresholding technique for this purpose, since it is more suitable for the detection of low-surface-brightness objects as opposed to peak finding which is better suited for stars. They make use of Lutz's one-pass algorithm [17] to extract eight connected contiguous pixels from a template frame, which is obtained by convoluting the original image with an appropriate mask. At the end of this step, image regions whose pixel values are above the computed threshold are considered potential object locations.

The deblending phase seeks to separate neighbouring objects that have been extracted as a single source. A series of 30 threshold levels, ranging from the primary extraction threshold to the peak value of the region, are applied to each group of connected pixels. A model of the light distribution of the

---

[3] At the time of writing, the original paper has collected over 7500 citations, with over 2500 since 2015, pointing to its widespread use in the astronomical community almost three decades after it was originally proposed.

object is obtained and stored as a tree structure as shown in Figure 4. The algorithm goes in a top-down fashion from the tips of the branches and at each junction decides whether it should extract two (or more) objects or continue its way downwards. To reallocate the pixels that have a flux lower than the separation threshold to one of the newly created subcomponents, the software uses a probabilistic approach where a bivariate Gaussian fit is used to estimate the likelihood of each pixel belonging to one of the new objects.

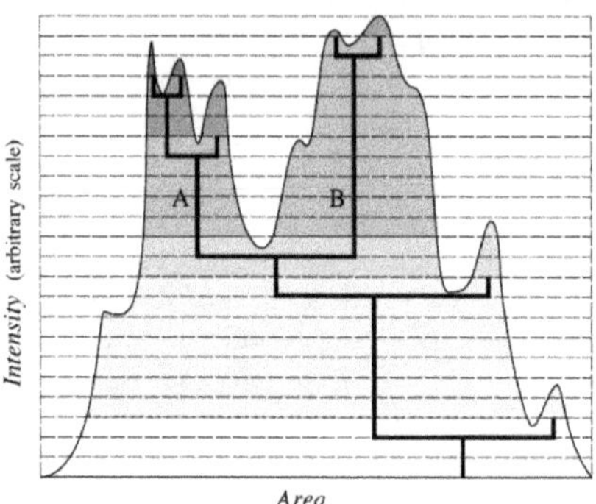

Figure 4 – Schema of the diagram used by SExtractor to deblend a composite object. The thick lines describe the profile of the object as a tree-structure. The decision to consider a branch as a distinct object is determined according to the relative intensity (tinted area). For this particular example, two components, A and B, are derived from the original object, with the pixels lying between the separation threshold being assigned to the most likely originator afterwards [16].

To avoid spurious detections of objects with bright profiles outside their centre, e.g., elliptical galaxies, SExtractor employs a filtering step. Since the background values in these regions are locally higher, a lower relative threshold is obtained and thus a higher change of detecting noisy peaks. To avoid this problem, the software verifies if there could have been a detection if neighbours of this region were not present. If, after the subtraction of neighbouring pixels, a detection would still be made, then the object is accepted as a detection, otherwise it is considered noise.

For the photometry measurements step, the package is capable of estimating isophotal, circular and total magnitude of an object. For each object, two types of total magnitude are computed, one using adaptive aperture, and the other using an isophotal correction. In the first approach, the second-order moments of the object profile are used to define an equivalent bivariate Gaussian profile with a mean standard deviation σ_{iso}. This is used to determine a shape within where the "first moment" can be calculated using equation (2.6) where r is the radius to a given pixel and $I(r)$ its pixel value:

$$r_1 = \frac{\sum rI(r)}{\sum I(r)} \tag{2.6}$$

The authors note that for stars and galaxy profiles convolved with Gaussian seeing[4], an almost constant flux is expected to lie within a circular aperture of $2r_1$, independently of their magnitude. For the corrected isophotal magnitudes, with the assumption that the intensity profiles of faint objects are approximately Gaussian, then the fraction of the total flux enclosed within a particular isophote, $\eta = I_{iso}/I_{tot}$, is given in equation (2.7):

---

[4] Seeing in this context is the equivalent of the PSF for ground-based observations where atmospheric turbulence degradation is present.

$$\left(1 - \frac{1}{\eta}\right)\ln(1 - \eta) = \frac{At}{I_{iso}} \tag{2.7}$$

where A is the area, and t is the threshold of the isophote. Although the above equation is not analytically invertible, a good approximation is obtained with the second-order polynomial fit given in equation (2.8):

$$\eta \approx 1 - 0.1961\frac{At}{I_{iso}} - 0.7512\left(\frac{At}{I_{iso}}\right)^2 \tag{2.8}$$

Once the fraction of the flux η is known, equation (2.9), can be used to estimate the total magnitude:

$$m_{tot} = m_{iso} + 2.5 \log \eta \tag{2.9}$$

Finally, for the star-galaxy separation step, the authors propose a neural network classifier. The architecture consists of the input layer, one hidden layer and a single output neuron (for binary star-galaxy classification) and uses the sigmoid as the activation function. A learning rate of 0.1 with a momentum parameter of 0.2 is also described. A total of 10 input features are proposed, with 8 corresponding to isophotal areas, one peak intensity and a final feature related to seeing.

They conclude with the experimental results obtained in both simulated and real images, where the SExtractor "appears to be reliable at least at a conservative $\approx 95\%$ level" (unclear is accuracy is meant), if heavily saturated stars are excluded, on both real and synthetic test sets.

2.1.3. OTHER COMMONLY USED SOURCE DETECTION METHODS

Although not as widely adopted as the methods described in Chapters 2.1.1 and 2.1.2, other commonly used approaches for the detection of astronomical sources make use of software such as DoPHOT, Starfinder, HSTphot (including its updated version DOLPHOT) and SDSS [2, 18].

DoPHOT is a software tool designed to search for astronomical objects on digital images and extract positions, magnitudes, and crude classifications for each. Its design is geared towards star / galaxy classification and stellar photometry using poorly sampled PSFs [19].One of its central features is its capacity to apply a bespoke model to each type of cosmological object that is being analysed. For detecting a star, the program may use an elliptical Gaussian, for a galaxy, a Gaussian may still be used albeit significantly larger; for a double star system, the model may comprise of two single stars These models are specified in terms of analytical functions with six parameters: x and y positions within the image, total intensity, and three additional shape parameters (height, wight and tilt). The program uses the initial seeing estimate to identify stars brighter than some high initial detection threshold. The analytical function used to represent a star is then fit to several different objects to determine a better estimate for the shape of a typical star or establish if the object is better represented by a non-stellar model (e.g., galaxy or cosmic ray). When an adequate fit is found, the respective object is subtracted from the image. The remaining objects are searched for successively fainter stars by lowering the detection threshold. After each pass through the image, all objects that have been previously removed are once again added to derive improved estimates of the models' parameters until a final catalogue with all the detections is produced.

The Starfinder algorithm [20] was specifically developed to analyse stellar fields and has a robust performance even in low-quality images. It extracts the PSF directly from the image frame to adequately account for the characteristics of the acquisition instrument and atmospheric perturbations. The first step consists of the PSF extraction, where the user starts by selecting a group of well-separated stars from a region of the image containing reduced noise. Several operations are performed to remove the background, normalize the pixel values, and smooth the selected shape to generate a representative template to be used across the frame. A similar technique to that described in SExtractor (Chapter 2.1.2) is used for background estimation and subtraction. For the phase of star detection and photometric measurements, the algorithm starts with a list of possible star shapes where the measured intensity subtracted from the background noise is larger than a predefined threshold. Starting with the higher intensity object from the prepopulated list, the program removes the star from the frame and recalculates if any new detections would still be made, compares the PSF shape with the precomputed template, and performs photometric analysis on the source object. Once every object in the list has been processed, a final re-fitting is performed to improve the photometric performance, and the background estimation updated. The process can be repeated if new stars are detected after the subtraction of the previously detected ones, a technique that is very useful in crowded stellar fields.

The HSTphot software was developed to assist in processing the vast sets of data generated by the Hubble Space Telescope (HST). Based on PSF-fitting stellar photometry, the program is not significantly different from DAOPHOT (see Chapter 2.1.1) and DoPHOT (see Chapter 2.1.2). It can be divided into five main steps: image preparation, PSF determination, detection of stars, iterative photometry solution, and aperture corrections [21]. The first step of image preparation is responsible for running important preliminary steps such as masking bad columns, cosmic ray cleaning, hot pixel masking, etc. The remaining unmasked and unsaturated pixels are scaled for their respective exposure times. For the following background determination phase, the sky value is calculated at each pixel using the mean of pixel values inside a square annulus centred on that point. The star brightness and position measurements are made by determining the combination of those parameters that best fit the observed data. The program then calculates a synthetic PSF which is then modified and magnified to compensate for any geometric or distortions errors. Once the sky background is subtracted from the image, the residuals are scanned for peaks. A series of passes are made, first scanning for the brightest pixels, then for slightly dimmer ones, iteratively. For any detected peak, using its centroid as an initial estimate, a photometry solution is run to improve the position and determine the star's brightness. After convergence, if the signal-to-noise ratio of the star exceeds a user-defined criterion, the star is kept, otherwise, it is rejected. Since the sky level is determined near the star's position, it will invariably contain some embedded starlight. To correct for this, the PSFs are also adjusted by subtracting the mean in the same region from all PSF values during the photometry solution. After the entire image has been processed to find stars of all brightness levels, all the detected stars are subtracted from the residual image and the process is repeated, allowing stars located in the wings of other brighter stars to be located.

The Sloan Digital Sky Survey (SDSS) is a large astronomical project consisting of four major components: a dedicated 2.5m telescope, a separate 0.5m telescope to monitor and provide calibration to the main equipment, a large format imaging camera, and a set of software tools for image analysis and processing. Two of the more prominent components of the computational package are the PSF estimation and star/galaxy separation tools [22]. The first step in PSF estimation is to identify a set of bright and isolated stars in the image. As with the aforementioned procedures, these are used to obtain the best estimate of what a representative PSF of a given frame would look like. However, the SDSS package doesn't use only a single frame of data to determine that frame's PSF, with stars from preceding and succeeding frames being also used. Some of the advantages are a better restriction of spatial variation at the edges of the frame, smoother variation of the PSF between frames, and an inclusion of more sources for the PSF estimation. Regarding the star/galaxy separation, three models are fit to every cosmological object in an image: a PSF, a pure de Vaucouleurs' profile (a relation between the surface brightness of a point on an elliptical galaxy and that point's distance from the galaxy's centre), and an exponential disk. These different models are employed to assist with the separation and morphological classification of the various cosmological objects, where the relative likelihood of their fit to the different functions is used to perform source separation and classification. However, the fit likelihoods were found to be particularly small for bright stars, due to the influence of slight errors in the modelling of the PSF. To correct this, the ratio of the flux in the best-fit galaxy model to that of the PSF was used as a discriminant.

2.1.4. NEW PROPOSALS AND ENHANCEMENTS IN SOURCE DETECTION PHOTOMETRY

Although some of the classical methods presented have seen widespread adoption and intense use by the astronomical community since their proposal, the field of stellar photometry has not been dormant after the turn of the century. A range of novel methods and variations have since been proposed that try to solve some of the previous limitations and provide the community with more robust and capable tools.

Metchev and Grindlay [23] propose a new algorithm for crowded field source detection of astronomical objects. The authors describe it as a *"two-dimensional version of the classical one-dimensional KS test"* (Kolmogorov–Smirnov test), where the test *"is used to optimize the goodness of fit in an iterative source-detection scheme for astronomical images"*. They aim to solve the problem that is encountered when the resulting PSFs of the individual point sources are heavily blended (which typically occurs when source separation <1.5 FWHM). In their proposal, provided that the PSF of the unresolved sources are known and constant over the image, a comparison between the image being analysed and a simulation of proposed source distributions can be used to obtain the KS probability that the image and simulation represent the same configuration of objects. Once the probability is obtained, it can be used as a measure of the accuracy of the positions and intensities of the proposed sources. Their proposal was compared with other source detection algorithms, such as DAOPHOT, having demonstrated superior power in heavily crowded fields with a low signal-to-noise ratio. Best fitting applications for this algorithm are in high-energy imaging, deep exposure of globular clusters and extragalactic nuclear regions, where the size of the PSF is typically in the same order of magnitude as the angle of separation between the objects.

Five years later, Magain et. al. [24] proposed a new method for crowded field photometry designed to determine the PSF of images that lack any bright or sufficiently isolated stars. To achieve this, they first approximate the PSF by either an analytical or a known numerical function (e.g., sum of Gaussians or Moffat). These functions are then fitted to all the point sources in the image, with the initial fit providing approximate values for the intensities and centres of the point sources are well as an approximate shape. The second step consists in adding a numerical component to the PSF estimate, done in a gradual process staring from the central region of the PSF outwards, to avoid the algorithm trying to fit the function to any existing bumps in the wings. The deconvolution problem is then: given an observed light distribution and an estimated deconvolution kernel, find the PSF of the deconvolved image. To solve it, the number of point sources, as well as initial estimates of their positions and intensities are then obtained using a standard algorithm (e.g., DAOPHOT). After the obtained residuals are inspected and unsatisfactory areas identified, additional point sources are added, and the process repeated until the fit becomes satisfactory. A comparison of the results obtained by DAOPHOT and this method in crowded astronomical fields, show the latter can obtain more precise and accurate measures with magnitude distribution errors reduced by a factor of 3.

A probabilistic Bayesian approach [14] has also been proposed for producing catalogues in crowded stellar fields. Brewer, Foreman-Mackey, and Hogg describe it as "capable of inferring the number of sources in the image and handling the challenges introduced by noise, overlapping sources, and an unknown point spread function. The luminosity function of the stars can also be inferred, even when the precise luminosity of each star is uncertain, via a hierarchical Bayesian model". The key idea of their proposal is that instead of computing a single catalogue for the image, their method constructs a posterior probability distribution of the space of possible catalogues that represent the current state of knowledge regarding the presence of stellar objects in each image. An example of the output is shown in Figure 5. With this, the uncertainties are accurately reflected in the scientific measurements that are drawn from the frame, e.g., concerning the luminosity function of stars. Regarding the results, the authors compared their proposed method with SExtractor (see 2.1.2), concluding that both achieved very similar and satisfactory performance for medium and bright stars, but their proposed approach was able to detect a larger number of faint stars. However, this is at the cost of increased computational time, with their Bayesian method being considerably more demanding and time-consuming than the former. They finally note some of the limitations of their approach, with the key remarks being the simplified representation of stars as depending only on position and flux, the lack of support for multi-band imaging and the difficulties in dealing with stellar clusters or situations where stars are not evenly spread out across an image.

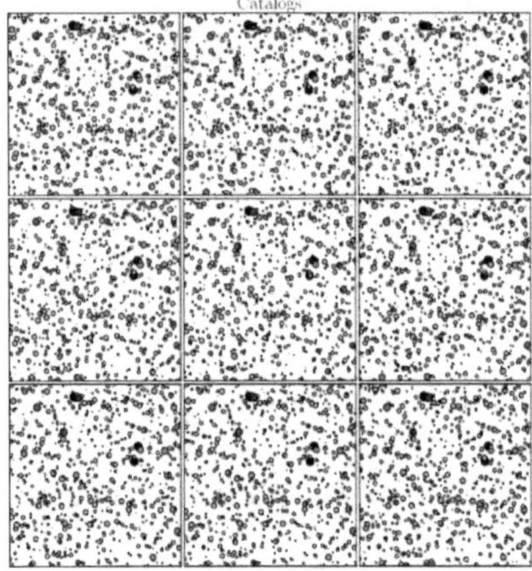

Figure 5 – Example output of the probabilistic Bayesian method consisting of samples of the distribution over the posterior space. Nine catalogues are shown sampled from the posterior distribution of one of the test cases. Common characteristics across them represent features with high probability, and differences represent regions of uncertainty. The area of each circle is proportional to the flux of each underlying star.

2.1.5. Shortcomings of classical source detection tools

Although the classical methods for source detection and the proposed variations and improvements have been extremely valuable for scientists and astronomers, they still present a number of challenges that new developments could address. Conventional source-finding algorithms present a limited ability to extract adequate point sources in extremely crowded star fields due to the relative size of the PSF compared to the angular separation of the underlying objects [23]. The authors further add that to find the signal-to-noise ratio, classical detection algorithms use either the average background (determined from a source-free region of the image) or local background (determined from a region around the detection cell). However, regardless of the approach, these techniques fail to distinguish between faint blended sources in crowded fields, where the background is affected by overlapping PSFs. Even though more recent attempts have been made at detecting increasingly fainter point sources [25, 26], an improved method that consistently improved the detection capabilities of astronomical software would allow for further scientific breakthroughs in coming studies.

Classical detection software shows difficulties in dealing with some challenging situations [14]. When multiple sources overlap, either partially or completely, these algorithms struggle to discern how many sources are present and what contribution of the flux should be attributed to each. A further, yet related, problem is raised by [24] concerning the determination of the PSF. In such crowded stellar fields, it is often the case that no star is sufficiently isolated from its neighbours to serve as the ideal candidate for PSF estimation. This may severely affect the photometric precision of traditional methods since any errors in the shape of the PSF will translate to photometric errors for these blended stars. They can only be accurately separated if the function used in the fitting process is the correct one.

Not only precision problems affect the classical detection methods. The Taiwanese American Occultation Survey (TAOS) [27] faced difficulties in performing real-time analysis when resorting to these conventional tools. SExtractor limitations are tied with the limited performance in low signal-to-noise ratio conditions, while DAOPHOT's slow computational speed greatly impacts its use.

2.2. OBJECT DETECTION AND INSTANCE SEGMENTATION IN DEEP LEARNING

Parallel to the applications of PSF estimation and model fitting for the detection of astronomical objects, a new field called computer vision, started emerging at the turn of the century, as detailed in Figure 6. It brought new techniques which sought to offer more generalized detection capabilities that could be applied to a broad spectrum of real-world problems [28].

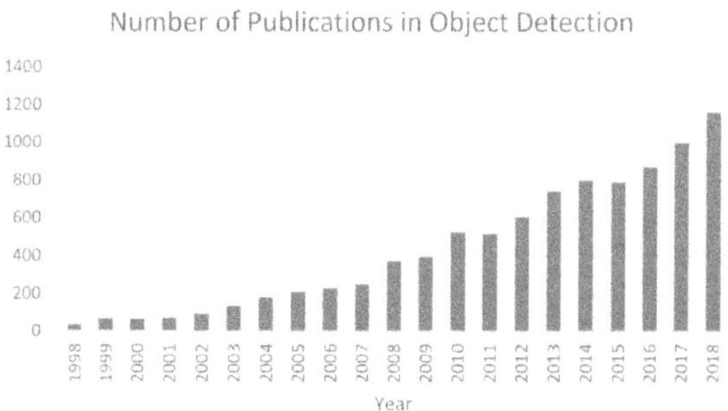

Figure 6 – Yearly publications in object detection between 1998 and 2018. Data was collected from Google scholar with the search terms "object detection" and "detecting objects" [28].

At the turn of the century, what became known as the "Viola-Jones detector" was originally proposed [29]. It was the first computer vision algorithm to achieve real-time detection of human faces without any constraints. This method used sliding windows to go through all possible locations in an image and detect if any contained a human face. The Adaboost algorithm was employed to select, from a large pool of random features, a small set of those that were deemed useful characteristics of a human face.

The Scale Invariant Feature Transform (SIFT) algorithm [30] is capable of extracting distinctive invariant features from an image that could be later used to detect the same object across different perspectives or in different positions within a scene. It starts by convolving the image with Gaussian filters at various scales. Difference-of-Gaussians (DoG) are generated from the previous convolution and candidate points of interest chosen from the maxima and minima of the DoG. The location of each candidate point is updated by interpolating the pixel values across pixels in the vicinity. Points situated in low contrast regions or along the edges are eliminated from the pool of candidates. Finally, orientations based on the peaks in the histograms of gradient directions are computed for the remaining candidates.

14

One year later the HOG detector was proposed [31]. The main idea behind this method is that the local appearance and shape of an object in a given image can be characterised in terms of the intensity distribution of gradients and direction of contours. These descriptors are obtained by dividing the image into small connected regions, called cells. A histogram of gradient directions or edge orientations is computed for each pixel in each cell, with their combination resulting in the descriptor. As shown in Figure 7, this method was originally proposed for the detection of pedestrians, although its use has since been expanded [32].

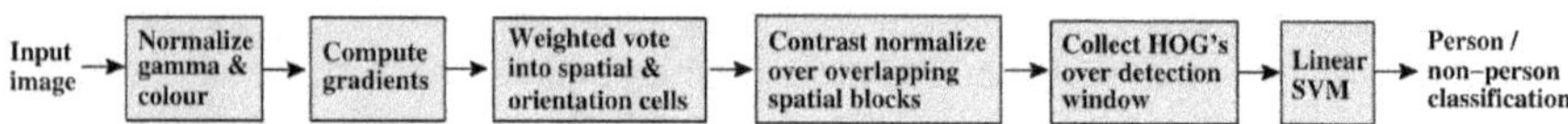

Figure 7 – Architecture overview of the HOG detector [31].

The popularity and performance of traditional object detection methods peaked in 2010 with the Deformable Part-based Model, or DPM [28]. Proposed as an extension of the HOG detector [33] its main idea was to break-up a complex task into a set of related subtasks that could more easily be completed. The training stage focused on finding an ideal way to decompose an object into its constituent parts and the inference stage on creating an ensemble of detections for the different object parts that could be used to determine the presence or absence of the original object.

As the 2010s arrived, the traditional methods started losing popularity to deep learning techniques, that were beginning to obtain promising results [34]. These new methods offered a key and very appellative advantage: instead of having to be tailored to a particular field and requiring bespoke adjustments and domain-specific knowledge, they could simply be given a dataset of annotated images and would be able to discover the underlying patterns present, learn the most descriptive and relevant features, and produce a prediction accordingly. Their downside was the additional computing and time resources consumed, but, as computing power kept growing and became more affordable, this shortcoming was quickly overcome [35].

2.2.1. REGION PROPOSAL-BASED FRAMEWORK

The first family of deep learning methods for object detection are region proposal-based frameworks, also known as two-stage detectors [36]. Their architecture can be divided in two main phases: the first, generation of proposals, seeks to find regions of interest that are bounded to contain a single object or feature within them; the second and final, class prediction, uses a convolutional neural network (CNN) based models to categorize the generated proposals into one of the predefined objects, or simply as background [37].

2.2.1.1. R-CNN

The introduction of R-CNN [38] which combined region proposals with CNN features was the first significant breakthrough in deep learning object detection methods. It significantly improved the quality of the generated bounding boxes during the region proposal phase and introduced a deep architecture to extract high-level features. When compared against the best performing method at the time (DPM), it was able to achieve improvements of around 30%, with a mean average precision (mAP) of 53% on the PASCAL Visual Object Classes Challenge 2012 [39]. Its architecture, illustrated in Figure 8, consists of three distinct phases. The first, region proposal generation, makes use of selective search to generate around 2000 region proposals for each input image. This method makes use of bottom-up grouping and saliency cues to rapidly determine candidate boxes and reduce the search space. In the second phase, during feature extraction, each region identified in the previous step is warped and cropped into a fixed resolution. The CNN then extracts a 4096-dimensional feature as a final representation that can be used for classification in the final stage. The last part, classification and localization, consists in applying a pre-trained linear Support Vector Machine (SVM) classifier to the extracted feature vectors. The different regions are classified according to the object detected or as being background. Finally, the positive regions (where objects are present) are adjusted with bounding box regression and filtered using greedy non-maximum suppression (NMS) to produce the final optimized bounding boxes corresponding to the location of the identified objects.

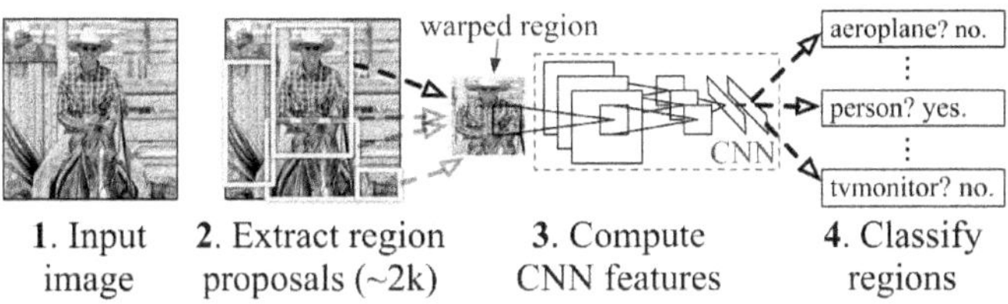

Figure 8 – Overview of the R-CNN architecture [38]

2.2.1.2. SPP-Net

The Spatial Pyramid Pooling (SPP) Network [40] sought to resolve the problem of having to input a fixed-size image, which often required cropping or warping that could impair the ability of the model to adequately recognize objects. To achieve it, the authors introduce a pooling strategy that allows the SPP network to generate fixed-size representations irrespective of the input size.

Detailed in Figure 9, the SPP's architecture starts with a series of convolutional layers that can accept an input image of any size and shape. These initial layers are responsible for extracting low-level features from the given frame. The innovative solution occurs at the following step. There, the spatial pyramid pooling layer takes the features given by the convolution process, which can be of any size or shape, and pools (or aggregates) the information into a fixed size vector that the final fully connected layers adopt for classification. The authors highlight the fact that this strategy aims to emulate the way the visual sense of the human brain is likely to work, by taking in and processing the entire image, producing a representation of the most relevant features and key points (convolutional step), aggregating the deeply processed information to some standard representation, and finally classifying the objects present.

The SPP network was able to achieve significant improvements in terms of speed and performance over R-CNN; however, not without its shortcomings. It features a very similar architecture to the R-CNN with the addition of its novel layer, thus requiring additional storage resources, but more importantly, the convolutional layers upstream of the SPP layer cannot be easily updated with the algorithm proposed, resulting in a loss of performance for deeper architectures [36].

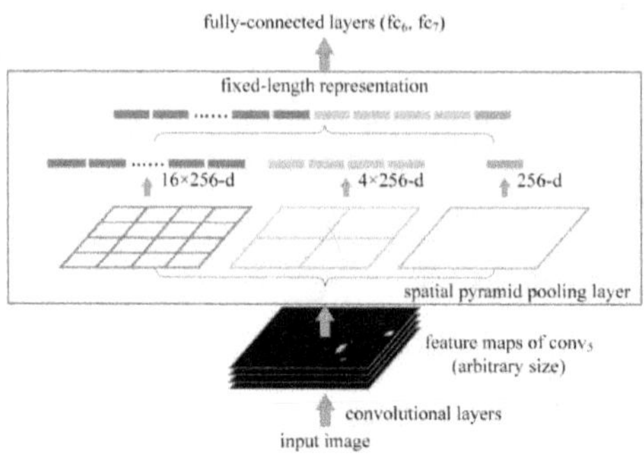

Figure 9 – Architecture of the SPP network [40]

2.2.1.3. Fast R-CNN

Girshick [41] noted a severe drawback with the two most popular object detection methods at the time, R-CNN and SPP. In the first, the author highlights the high spatial and time costs due to features being extracted for each object proposal within each region and stored in disk as well as the slow detection process of 47 seconds per image. For the second, storage limitations during data processing and speed of deep networks were also noted.

To address these, an improved version of the R-CNN, named Fast R-CNN, was proposed [41]. With its architecture illustrated in Figure 10, this method receives an image and processes it with several convolutional and pooling layers, to produce a vector of the relevant features. Each feature vector represents a region of interest (RoI) and is passed onto a collection of fully connected layers, which in turn pass it to two output layers. A layer with a softmax activation function is responsible for classifying the region as either one of the classes, or a generic background. The other layer is given the task of determining four numerical values, one for each vertex of the box that bounds the object.

To build upon its predecessors and improve training efficiency and speed, Fast R-CNN makes use of stochastic gradient descent that hierarchically samples mini-batches for training. During each training step, it selects a given number of images and samples a set of regions of interest within them. If these RoIs originated from the same image, they can share computational resources in both forward and backward propagation. The labels of the true bounding box v and true class u are given to each RoI during the training phase, so that a multi-task loss L can be jointly computed for its outputs, as detailed in equation (2.10):

$$L(p, u, t^u, v) = L_{cls}(p, u) + \lambda \, [u \geq 1] L_{loc}(t^u, u) \tag{2.10}$$

where $L_{cls}(p, u) = -\log p_u$ defines the loss assigned to the ground truth class and $L_{loc}(t^u, v)$ is calculated from the predicted t^u and true v coordinates of the respective bounding box.

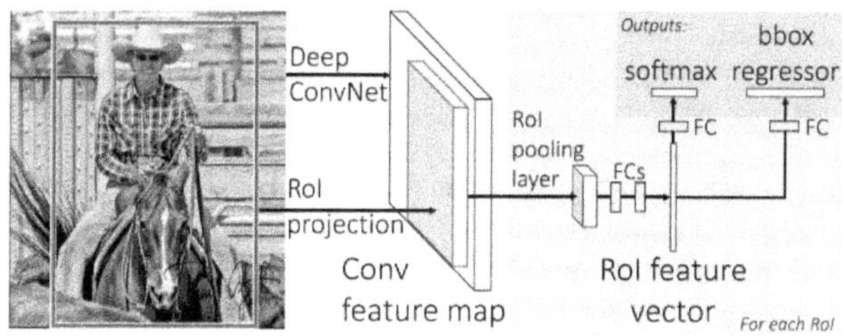

Figure 10 – Architecture of the Fast R-CNN [41]

2.2.1.4. Faster R-CNN

A further improvement to the R-CNN family of models came two years later with the proposal of Faster R-CNN [42]. Despite the success of previous architectures, the authors note that some of their shortcomings were still to overcome. One of the key drawbacks of the original proposal [38] was the large computational overhead that made real-time predictions hardly possible. Fast R-CNN solved this problem by sharing convolutions across region proposals and can achieve near real-time speeds even for very deep networks (if the time spent proposing regions isn't included). The inefficiencies still to resolve were the reliance on methods such as EdgeBox (which represents an image by its contours and the count of contours within a given bounding box is related to the likelihood of it containing an object) or the slow speeds of computing the region proposal.

Faster R-CNN's solution for these shortcomings, illustrated in Figure 11, was to introduce a new region proposal network (RPN) that can achieve computationally inexpensive proposals, allowing it to run at near real-time speeds, by sharing image-wide convolutional features between the RPN and detection stages. The RPN [42] consists of a fully convolutional network that can simultaneously determine a set of bounding boxes around an object and propose a score for each location. It further simplifies the detection process by sharing computations with the detection layers of the Fast R-CNN model.

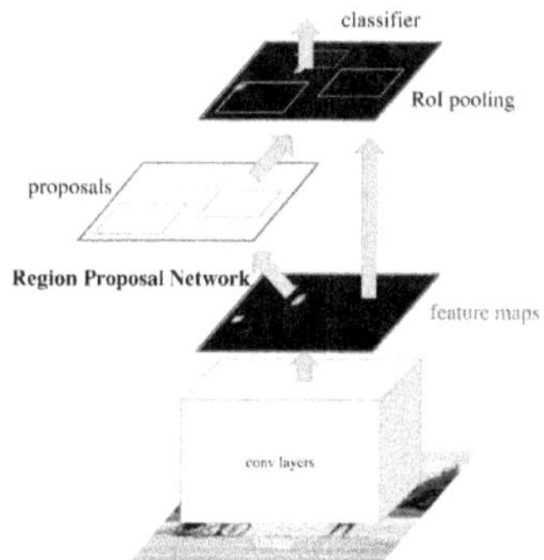

Figure 11 – Overview of the architecture proposed for the Faster R-CNN [42]

The region proposal step operates by sliding a small network over the feature map that is outputted by the convolutional layer. As visible in Figure 12 (which denotes the arrangement of this network at a single position), the features extracted by the sliding window are converted to a lower-dimensional space (256 dimensions in this case) from where they can be passed onto a *cls* and a *reg* layer. The first performs box classification, i.e., determines whether an object is present, and the second performs bounding box regression, i.e., determines the coordinates of the surrounding box. Additionally, a set of predefined anchors are used at each position of the sliding window to determine which shape and scale better describes the potential object region. The loss function used to adjust the regression of the predicted and real bounding box positions is detailed in equation (2.11):

$$L(p_i, t_i) = \frac{1}{N_{cls}} \sum_i L_{cls}(p_i, p_i^*) + \lambda \frac{1}{N_{reg}} \sum_i p_i^* L_{reg}(t_i, t_i^*) \qquad (2.11)$$

where p_i represents the probability that anchor i contains an object, p_i^* takes the value 1 for a positive anchor and 0 for a negative one, t_i denotes the four coordinates of the proposed bounding box and t_i^* corresponds to the known bounding box of the object. Finally, L_{cls} and L_{reg} correspond to a logarithmic and regularized loss, which are normalized by the size of the minibatch used, denoted by N_{cls}, and the number of anchors used N_{reg}, respectively.

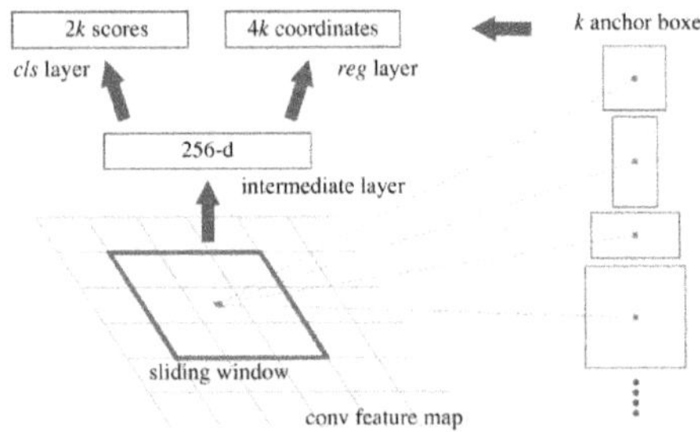

Figure 12 – Overview of the design adopted for the region proposal network [42]

2.2.1.5. FPN

Feature pyramids have been a well-known and researched area in object detection and were the precursor to some successful methods in the mid-2010s [40, 43], excelling in the detection of objects at varying scales. However, their high computational costs have been a detractor for wider adoption, both in the industry and academia. Further research into this type of algorithms eventually lead to the proposal of Feature Pyramid Networks (FTP) [44], which promised to take advantage of the pyramidal hierarchy and their scale-invariant property (i.e., objects at different scales in an image can be adequately represented at different levels in the pyramid) to provide robust object detection capabilities with marginal computational cost.

The FPN approach, illustrated in Figure 13, consist of a bottom-up and a top-down pathway, the latter of which is enhanced by lateral connections. The bottom-up part of the network takes advantage of the pyramidal property of CNNs (typically provided by the convolutional and pooling steps) to construct representations of the input images spanning from low-level detailed features (at the bottom levels) to more general and abstract concepts (at the higher levels). The top-down path is then constructed by upsampling the higher layers of the pyramid to generate lower levels with higher resolution features. At the same time, lateral connections to the bottom-up part are introduced to join feature maps from equally sized levels. Despite containing lower-level semantics, they can still provide more precise activation maps since they accumulate fewer subsamples.

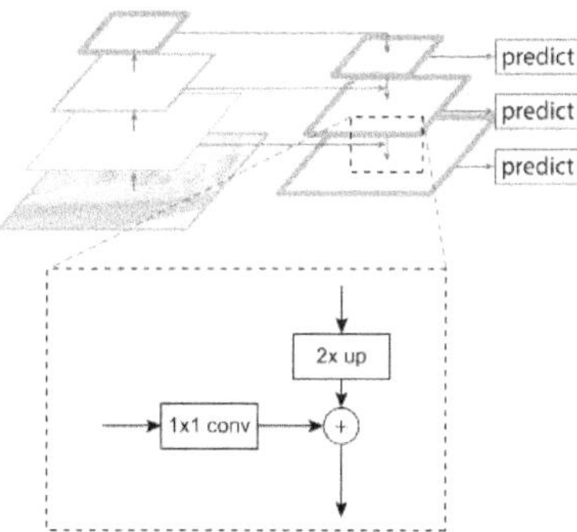

Figure 13 – Overview of the FPN architecture, comprised of bottom-up and top-down pathways, as well as lateral connections [44]

2.2.1.6. Mask R-CNN

Mask R-CNN [45] was conceived as an extension to the state-of-the-art method Faster R-CNN (see Chapter 2.2.1.4), with the ability to perform instance segmentation. Instance segmentation aims not only at localizing an object in an image (object detection), but also at providing a classification for the object and differentiating between instances of the same type. Mask R-CNN proposed to achieve this by adding a third branch to the Faster R-CNN architecture, responsible for predicting a binary mask associated with each RoI. The architecture of this model is illustrated in Figure 14.

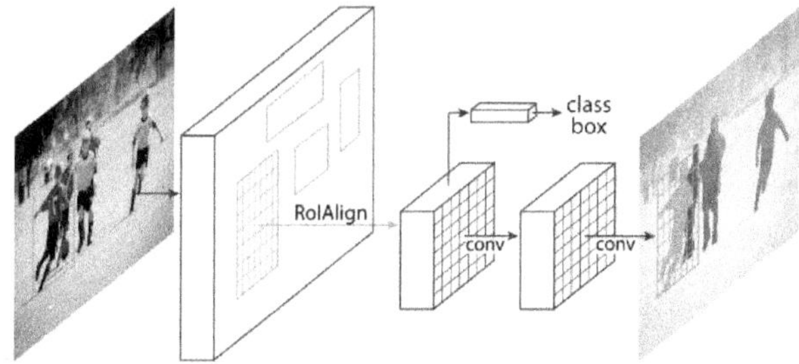

Figure 14 – Representation of the Mask R-CNN framework [45]

Unlike the two output branches originating from Faster R-CNN, which are comprised of fully connected layers that collapse a two-dimensional region into a one-dimensional vectorial representation (inevitably losing information in the process), the additional segmentation branch uses a square mask to maintain the spatial arrangement of the original object, as illustrated in Figure 15. This convolutional layer can produce a representation of the region with a smaller number of parameters while still maintaining strong accuracy. With the addition of the third branch, the loss function was updated to account for the segmentation task. This is done by expanding equation (2.11) with the addition of a mask loss L_{mask}, which is calculated by applying a sigmoid function to each pixel within a RoI and taking the binary cross-entropy loss across all regions of each class.

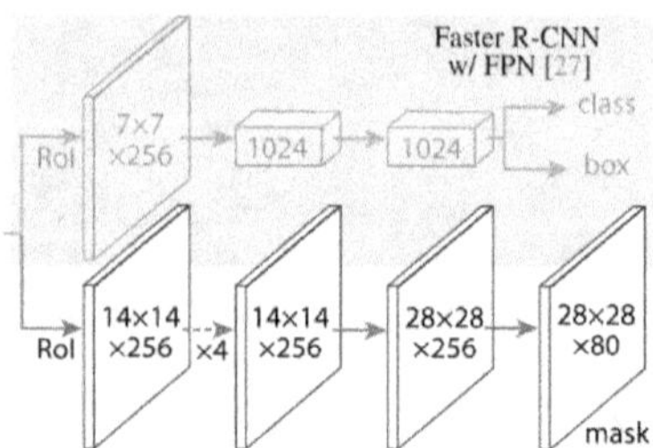

Figure 15 – Segmentation convolutional branch of Mask R-CNN for a Faster R-CNN with FPN architecture [45]

A further difference between Faster R-CNN and Mask R-CNN relates to the way each extracts feature maps from the RoIs. Faster R-CNN uses the RoIPool technique, based on quantization, which is susceptible to introducing misalignments between the extracted features and the underlying RoIs. Although classification is robust to small alignment errors, when the task is predicting masks that are accurate to the pixel level, that is no longer the case. To overcome this problem, Mask R-CNN introduces RoIAlign. This technique overcomes the quantization problems of the former by using bilinear interpolation to calculate the precise values of the input features at four equally spaced positions in each region.

2.2.2. REGRESSION/CLASSIFICATION BASED FRAMEWORK

Although region proposal-based frameworks have been able to achieve very robust results in object detection and classification, their two-step process of region extraction and object classification with further localization refinement still poses significant computational and time overhead, particularly for applications that require real-time results. To address these limitations, a different set of methods known as regression/classification-based frameworks or one-stage detectors, were developed. They aim to apply global regression/classification to map features directly from the pixels of the input image to the prediction of both the bounding box and the class of the object.

2.2.2.1. YOLO

The You Only Look Once (YOLO) algorithm [46] was the first single-stage detector to achieve a robust performance in the era of deep learning [37]. Its architecture, illustrated in Figure 16, is significantly simpler than that of methods discussed in Chapter 2.2.1. The process starts with the image being divided into a $S \times S$ square grid. Each square, or cell, is given the responsibility of detecting the object that falls within its borders. Cells do this by defining a series of bounding boxes within their borders for where the object may be and assigning a confidence score to each. The scores represent the confidence the model has that an object is truly bounded by the proposed box. More formally, this confidence can be defined as $\Pr(Object) * IOU_{pred}^{truth}$ where the first term represents the likelihood that an object exits, and the second the intersection over union (IOU) of any real box and predicted bounding box.

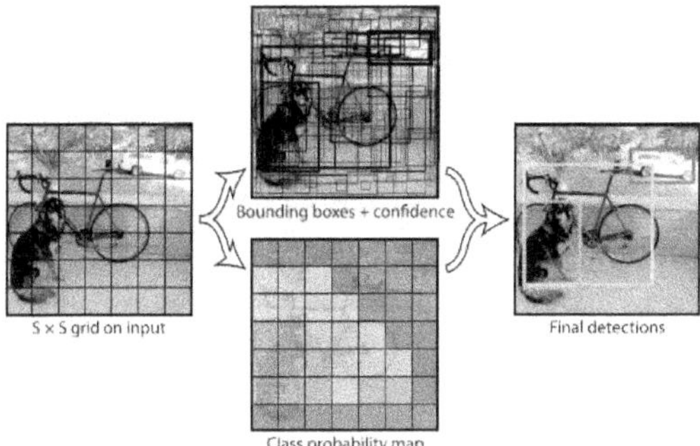

Figure 16 – Overview of the YOLO detection method [46].

Regardless of the number of bounding boxes proposed at each cell, a single conditional probability is also estimated for each class that the object may belong to Pr $(Class_i|Object)$, provided that the cell is thought to contain an object. During the test phase, the confidence scores for each of the possible classes can be obtained by multiplying the confidence that the bounding box contains an object by the conditional probability of that object being of class i, as per equation (2.12):

$$\Pr(Class_i|Object) * \Pr(Object) * IOU_{pred}^{truth} = \Pr(Class_i) * IOU_{pred}^{truth} \tag{2.12}$$

The architecture of the YOLO network, detailed in Figure 17, is comprised of 24 convolutional layers situated upstream of two fully connected layers. Based on [47], 1×1 reduction layers are used, followed by convolutional layers of size 3×3. The network can work with real-time images reaching speed of 45 frames per second while still being able to achieve a robust performance when compared with other state-of-the-art detectors. Despite being more prone to localization errors, it typically predicts fewer false positives, i.e., classifying the background as an object. Subsequent improvements to this model have since been proposed with YOLOv2 [48], YOLOv3 [49], YOLOv4 [50], although they largely follow the principles discussed with various optimizations in both accuracy and speed.

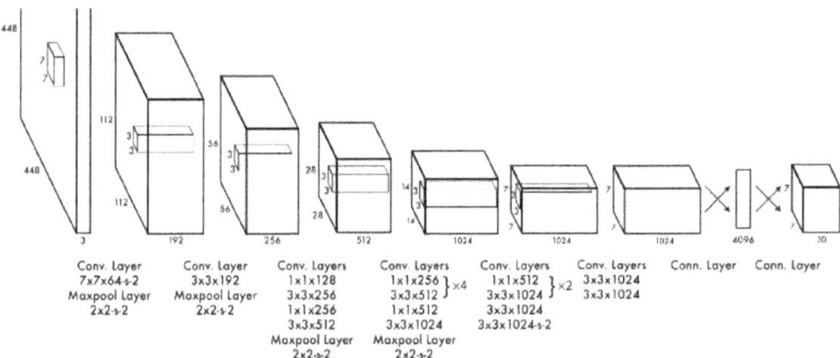

Figure 17 – Architecture of the YOLO network [46]

2.2.2.2. SSD

Another popular one-stage detector model was proposed in the same year as YOLO, and named SSD, or single-shot detector [51]. It aimed to address some of the shortcomings of YOLO, namely its difficulties in performing accurate localizations (when compared with cutting edge region proposal-based detectors) and its inability to recognize more than two objects at a given location, making it inadequate for scenes with small or crowded objects [37].

SSD uses a base model, namely VGG16 [52], which is typically employed in tasks of high-quality image classification, to make up the initial layers of its network, as detailed in Figure 18. Not the entire VGG16 model is used, but only the initial layers responsible for the convolutional and pooling steps, with the final fully connected layers responsible for classification being discarded. From there, the SDD model starts by adding convolutional feature layers that progressively decrease in size and allow for the network to detect objects at multiple scales. Each of these makes use of a 3×3 kernel that outputs either a category score or a relative shape offset from the default bounding box position. Unlike YOLO, this model doesn't start with a standard gird overlayed onto the image, but instead a collection of predefined bounding boxes with varying aspect ratios and scales are used to construct the bounding boxes. Training is done by minimizing a loss function similar to equation (2.10), consisting of a weighted sum of a confidence loss via the use of the softmax function, and a localization loss through the smooth L1 loss.

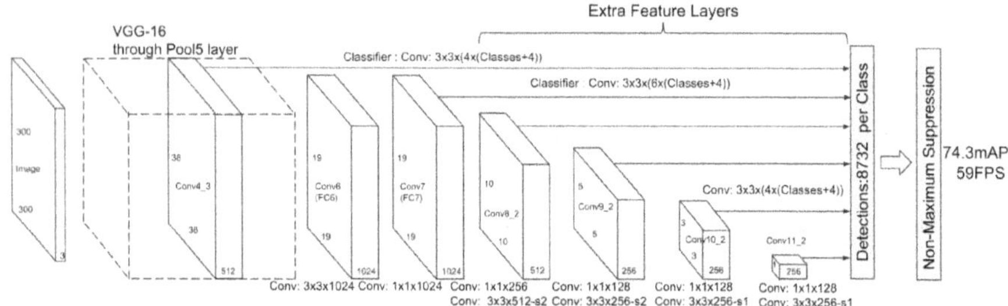

Figure 18 – Architecture of the SDD model [51].

Additional features are also included in SDD, such as selection of default box scales and aspect ratios, hard negative mining, and data augmentation. All these considerations allow it to have a very robust performance, even surpassing the accuracy of the state-of-the-art two-stage detector Faster R-CNN while maintaining speeds nearly three times as fast. However, some drawbacks are still present, such as the difficulty in keeping robust performance levels when detecting small objects [36].

2.2.3. PERFORMANCE COMPARISON

A comparative study of different object detection methods [36] looked at their performance in some of the most popular and commonly used benchmarking datasets, namely PASCAL VOC 2007 [53], PASCAL VOC 2012 [39] and Microsoft COCO [54]. The first PASCAL dataset contains 9 963 images of 20 distinct classes (including persons, animals, vehicles, and indoor objects), totalling 24 640 objects. PASCAL VOC 2012 maintains the same number and type of classes but adds some improvements, such as a larger total number of images, now 11 530, 27 450 RoI annotations and 6 929 segmentations. The Microsoft COCO dataset is much larger, containing photos of 91 objects, with a total of 2.5 million labelled instances and 328 000 images. The authors have used VGG16 [52] pre-trained on ImageNet [55] as a base model when no other was specified by the authors of the original networks and mAP was the selected metric used for comparison. The results obtained are detailed in Table 1 for the performance using the test set of each of the aforementioned datasets.

Table 1 – Summary of the mAP and AP score obtained by the different object detection algorithms over different datasets. Adapted from [36].

Method	PASCAL VOC 2007		PASCAL VOC 2012		Microsoft COCO		
	Trained on	mAP	Trained on	mAP	Trained on	mAP [0.5:0.95]	AP (small object)
R-CNN	07	66.0	12	62.4	No data available		
SPP-net	07	60.9	No data available		No data available		
Fast R-CNN	07+12	70.0	07++12	68.4	train	20.5	4.1
Faster R-CNN	07+12+COCO	78.8	07++12++COCO	75.9	trainval	24.2	7.7
FPN	No data available						
Mask R-CNN	No data available				trainval35k	**39.8**	**22.1**
YOLO	No data available		07++12	57.9	No data available		
YOLOv2	No data available		07++12++COCO	78.2	trainval35k	21.6	7.7
SSD300	07+12+COCO	79.6	07++12++COCO	79.3	trainval35k	23.2	5.3
SSD512	07+12+COCO	**81.6**	07++12++COCO	**82.2**	trainval35k	26.8	9.0

Regarding the notation of Table 1 – Summary of the mAP and AP score obtained by the different object detection algorithms over different datasets. Adapted from [36]., the following applies:

- AP: average precision for an IOU of 0.5

- mAP: mean average precision for an IOU of 0.5

- mAP[0.5:0.95]: average of the mean average precision calculated for a range of IOUs between 0.5 and 0.95 at 0.05 interval size

- 07: training on the PASCAL VOC 2007 train and validation data

- 12: training on the PASCAL VOC 2012 train and validation data

- 07+12: training on the union of training and validation data of PASCAL VOC 2007 and 2012

- 07+12+COCO: training on 35 thousand samples of the COCO train set and fine-tuning 07+12

- 07++12: training on the union of PASCAL VOC 2007 train, validation and test data with PASCAL VOC 2012 train and validation data

- 07++12++COCO: training on 35 thousand samples of the COCO train set and fine-tuning 07++12

Although one-stage detectors such as SDD have a good performance on the simpler PASCAL VOC 2007 and 2012 datasets, when exposed to the more complex Microsoft COCO image data, the two-stage detectors, led by Mask R-CNN perform considerably better. The authors suggest this is due to the classification/regression methods struggling with accurate object localization. They further add that one of the key reasons that explain why the performance is considerably lower on the Microsoft COCO is due to the larger number of small and nonstandard objects.

One year later, two further studies [56] and [37] looked at the performance of the various published methods on both the PASCAL and Microsoft COCO datasets. The reported results are summarized in Table 2 and Table 3, respectively. On the richer Microsoft COCO dataset, Mask R-CNN shows a slight lead in performance when compared to the Faster R-CNN implementation, including in small objects. Its ability to extract the object mask, instead of the bounding box leads to fewer perturbations from neighbouring objects that can be unintentionally included for later classification tasks. Regarding the comparison between one-stage and two-stage detectors, although YOLOv2 and SDD512 were able to obtain slightly better results for the simpler PASCAL dataset, the data from the Microsoft COCO experiments suggests that their performance may suffer for more complex tasks.

Table 2 – Summary of the mAP and AP score obtained by the different object detection algorithms over different datasets. Adapted from [56]

Method	PASCAL VOC 2007		Microsoft COCO		
	Trained on	mAP	Trained on	AP	AP (small objects)
R-CNN	07	59.2			
SPP-net	07	60.9		No data available	
Fast R-CNN	07+12	70			
Faster R-CNN	07+12	73.2	trainval35k	36.2	13.6
FPN			No data available		
Mask R-CNN	No data available		trainval35k	**39.8**	**22.1**
YOLO	07+12	66.4			
YOLOv2	07+12	**78.6**		No data available	
SSD300	07+12	74.3			
SSD512	07+12	76.8			

Table 3 – Summary of the mAP and AP score obtained by the different object detection algorithms over different datasets. Adapted from [37]

Method	PASCAL VOC 2007	PASCAL VOC 2012	Microsoft COCO	
	mAP	mAP	AP	AP (small objects)
R-CNN	66	62.4	No data available	
SPP-net	60.9	No data available		
Fast R-CNN	78.9	68.4	19.7	No data available
Faster R-CNN	69.9	70.4	36.2	18.2
FPN	No data available			
Mask R-CNN	No data available		**39.8**	**12.3**
YOLO	No data available	57.9		No data available
YOLOv2		78.2	21.6	5
SSD300	79.6	79.3	No data available	
SSD512	**81.6**	**82.2**	28.8	10.9

3
STATE OF THE ART

3
STATE OF THE ART

Large amounts of data are currently being generated by a range of astronomical surveys, e.g. [57, 58]. This high volume of information is expected to continue or further increase in the coming years, posing new challenges for scientists and astronomers that are now seeking better strategies to process and analyse all the new data [59].The popularity growth of deep learning methods for object detection and instance segmentation, largely powered by their remarkable performance improvement over recent years, offers a very compelling solution to this problem. It promises not only to be able to process multiple images per second but to do it with an accuracy and precision comparable to, or even surpassing that of the most highly trained humans.

This has led to increased adoption of machine and deep learning methods to perform once programmatic tasks, such as the detection of astronomical sources, their photometric analysis and classification. New algorithms have been developed over recent years to address the tasks of synthetic image generation, PSF estimation and detection and segmentation.

3.1. SYNTHETIC IMAGE GENERATION

Although vast amounts of data are being generated by satellites and capturing devices both on earth and in orbit, the information they collect is still largely unlabelled, i.e., the ground truth regarding the objects that generated an image is not known. To develop accurate tools and software that can analyse the resulting images to the high precision desired, these algorithms often need to be trained and tested with mock data that is as realistic as possible [60].

Peterson et. al. [61] proposed an astronomical image simulator named Photon Simulator, or PhoSim. It uses a Monte Carlo approach to create images by sampling photons originating from models of astronomical point sources and simulating the effect that their interaction with the atmosphere, telescope and camera have on the resulting image shape and characteristics. Several simplifications of the true underlying process are employed, without sacrificing the precision of the simulations, as the physical processes that define the location, shape and brightness of stars and galaxies can still be accurately represented with this approach.

Modelling the photon properties from the source until it arrives at a detector on Earth is done using ray optics. Although an electromagnetic or quantum-mechanical interpretation can be appropriate for certain interactions (e.g., diffraction), this approach is computationally more efficient yet still able to capture a representative behaviour. The photon is described by a set of four properties: a vector representing its position, a unit-vector characterising the angle of propagation, a time stamp measuring the time the photo arrived on Earth, and the wavelength of the photon. Alterations of the photon variables may be due to a change in trajectory, a change in position, or removal of the photon (e.g., due to scattering sending it in a different trajectory). A Monte Carlo model is used to describe the changes in trajectory, intensity, or throughput.

Diffuse emissions are considered via the airglow from the dark sky, the moonlight reflected and additional emissions near twilight. These types of diffuse light are modelled by a set of uniformly spaced sources on the celestial sphere that are set apart by 15 arcsec. The same spacing is used for the standard deviation of the Gaussian functions that model them, resulting in a statistically uniform illumination pattern. The sky simulation is completed by removing a portion of the photons to represent dust absorption.

The atmospheric structure is modelled via several parallel layers that cover the turbulence blurring and photon absorption within two altitudes. The modelling of atmospheric turbulence is done using the von Karman spectrum and a phase shift that is a function of the spatial position is used to represent the effect of turbulence on light propagation, caused by the temperature variations produced by airflow turbulence. The interaction of the photons with the turbulent atmosphere is characterized by the point spread function, and modelling of molecular opacity and atmospheric dispersion are also included.

Dome seeing may occur as the photon passes through the optical elements of the capturing system. This is modelled by disturbing the photon's trajectory using an isotropic Gaussian angular distribution. Diffraction is accounted for by including an Airy disk component for the photon's entrance in the pupil, and the scattering that occurs from the irregular mirror surfaces is captured using empirical methods.

Hot pixels are included in the image by selecting a random number of pixels and including enough electrons to simulate their saturation. Conversely, dead pixels are modelled by removing all electrons from them. Hot columns and dark current are also included in the model. Finally, digitization is approximated by equation (3.1):

$$ADU = \frac{e}{G(1 + N\frac{e}{W})} + B \tag{3.1}$$

where e accounts for the number of electrons in a pixel, G for the gain, W represents the full well depth, B the bias and N a non-linearity factor.

An illustration of how the various physical processes described herein are modelled and interact to form the final image is illustrated in Figure 19. Each step of the physical simulation of the various phenomena contributes to the final shape of the PSF, the position of the star and photometric intensity.

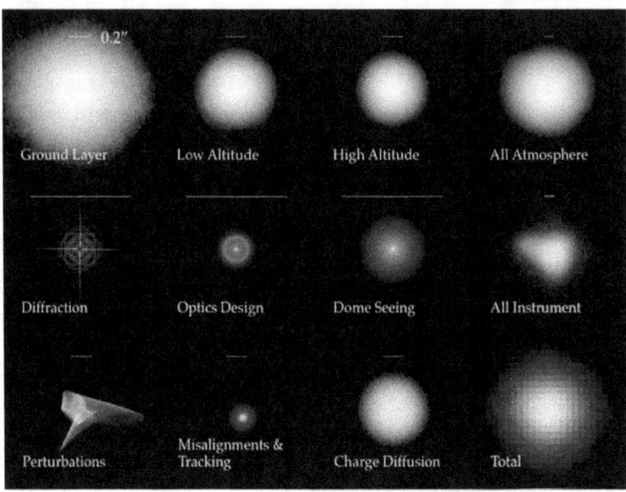

Figure 19 – Illustration of the various physical processes modelled by PhoSim to generate the image representation of a single star [61]

3.2. RECENT ADVANCEMENTS IN PSF ESTIMATION

Accurate estimation of the PSF has been a central problem for observational astronomy over the past decades. Regardless of the type of object being observed, i.e., galaxy, star, etc., the resulting image produced by these sources is described by the PSF that is caused by the imaging process (including atmosphere, telescope, instrument, and detector). This model can describe a broad spectrum of phenomena that may affect the captured shape of the emitting sources, such as the finite aperture of the telescope, charge diffusion of digital detectors, imperfections that may exist in optical equipment, and the turbulence caused by the Earth's atmosphere. As modern technology develops and is able to produce more accurate estimations of these functions, interest in novel techniques to model these phenomena has also been growing. [62] launched a challenge to reconstruct the original PSFs from collections of images of stars and galaxies. Although the focus was on the correction and reconstruction of galactical properties, the challenges for stars are similar. The authors concluded that the proposed methods struggled to model narrower PSFs adequately, and that the reconstruction was considerably impacted by the inclusion of atmospheric turbulence. Over recent years, new approaches have been proposed that formulate the estimation of the PSF as a nonparametric machine learning problem.

3.2.1. FAST PSF MODELLING

Motivated by the importance of accurate PSF estimation to study galactical properties and morphology, [63] proposed a deep learning-based model, that uses convolutional neural networks to estimate the parameters that characterize the PSF. This approach attempts to learn all the relevant features from each individual frame of data, instead of relying solely on known physical properties.

The first step involves learning the feature that better represents the model and ought to be captured. This is done via principal component analysis (PCA), which aims to generate a set of vectors that the PSF data may decompose into, and that can be ordered accounting for the amount of variation that each introduces in the data. To simulate a star where the PCA can be applied, a number of photons are sampled from a PSF profile computed by a weighted sum of two Moffat functions. Once this is known, a series of transformations are applied to allow for deviations from the base profile. The first transformation is to account for variability in the size of the PSF. The second attempts to capture its shape properties, such as skewness, ellipticity and kurtosis. The third and final set of transformations accounts for shifts to the centre of the profile.

With the basic features that may affect the final shape understood, a deep learning model is proposed, as illustrated in Figure 20, to learn which combination of parameters would produce an adequate model for a given star. It consists of various convolutional and pooling layers before converting the information from the lower-level features into a 1-dimensional representation consisting of 2048 fully connected neurons. From there, the final outputs can be generated as a regression problem, including the centre offset $\Delta\theta_i$, the FWHM, the ellipticity e_i, the skewness $\tilde{f}_i$, the triangularity $\tilde{g}_i$, and the kurtosis $\tilde{k}$.

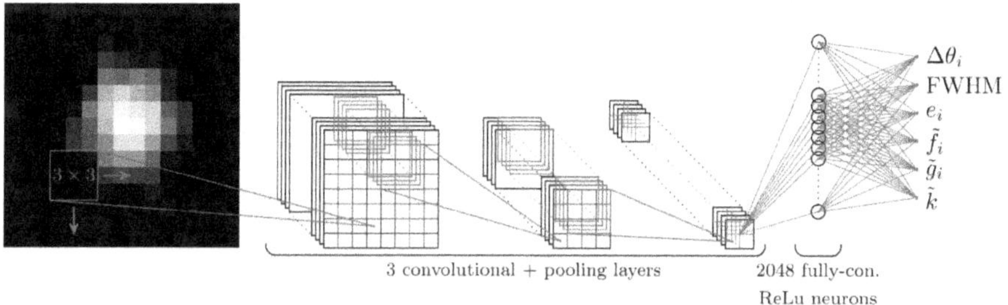

Figure 20 – Schema of the CNN used for the fast PSF modelling [63]

Comparing this solution to direct fitting, the authors found that although similar results were obtained performance-wise, the deep learning approach was able to perform considerably faster once training was completed. The traditional method requires several seconds per star, while the CNN can process thousands of stars per second, making it ideal for application in large surveys where vast amounts of data are collected.

3.2.2. ANALYTICAL PSF ESTIMATION FOR ADAPTIVE-OPTICS INSTRUMENTS

Initially proposed by [64] and later extended by [65] this method focuses on the application of analytical tools for PSF modelling in systems assisted by adaptive-optics. Being difficult to estimate in an accurate and systematic manner, due to phenomena such as crowded populations or low signal-to-noise ratio, the aim of this method is to be able to perform PSF reconstruction for any adaptive optics instrument, with a single and flexible algorithm.

Seven parameters are used to characterise the PSF estimation model: r_0 is a constraint applied to the uncorrected power spectral density corresponding to the PSF wings (i.e., the region that is unaffected by adaptive-optics corrections); C represents a constant used to change the gap between the part corrected by adaptive-optics and the uncorrected high-spatial frequencies; A describes the total energy contained in the Moffat power spectral density model; α_x and α_y are used to regulate the elongation and skewness of the power spectral density, while the angle θ governs the direction of the elongation; finally β represents an approximation of the PSD slopes for larger spatial frequencies within the adaptive-optics correction radius (typically between 1 and 1.83 for nominal correction, although higher values are possible for non-atmospheric aberrations).

The PSF fitting process consists in estimating a model of the image (including four additional parameters to adjust the PSF position, flux and background level), defining a criterion that minimises the squared distance between the estimated PSF and the real data PSF, and applying an iterative algorithm to perform the adjustment and optimization. Figure 21 illustrates different kinds of PSF reconstructions obtained for varying types of adaptive-optics systems. Despite some visible residuals, due in part to the presence of static aberrations or short exposure times, this modelling technique is able to accurately reproduce the observed data and create adequate reconstructions for different types of acquisition equipment.

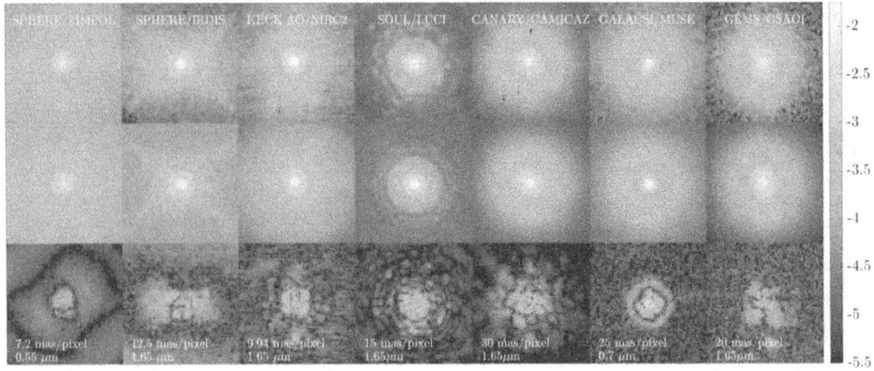

Figure 21 – Comparison of the real PSF (top), the PSF obtained via model fitting (middle) and the residuals between the two (bottom). Each column represents a different kind of adaptive optics system [65].

To evaluate the performance of the parsimonious model, a total of 4812 point spread function were collected from different types of imaging instruments across the planet, all sharing the use of adaptive optics as a common factor. Results showed that this technique was able to obtain a 0.7% bias and 4.0% standard deviation when estimating the Strehl ratio (SR – ratio between the peak intensity of an image divided by the peak intensity of a diffraction-limited image with the same total flux) and a bias of -0.8% with a standard deviation of 4.6% for the estimation of the FWHM of the PSFs. The reconstruction error obtained was of 4%, remaining consistent even for high SR images.

3.3. STELLAR DETECTION APPROACHES BASED ON DEEP LEARNING

Over recent years, developments in ease-of-use, accessibility and performance of deep learning techniques have led to an increasing number of studies and scientific publications on the use of these methods to study stars and other cosmological objects. Detection and classification of the different types of objects have been some of the more prominent tasks gaining growing attention from the scientific community.

3.3.1. ASTRONOMICAL TARGET DETECTION AND CLASSIFICATION FRAMEWORK BASED ON FAST R-CNN

Jia and Sun [66] proposed a method based on the Fast R-CNN architecture that both identifies the location of a source, and provides a classification of the type of object that has been detected. As highlighted in Figure 22, the proposed architecture relies on a backbone of ResNet50 used to extract the main features from the images that can then be used for subsequent region proposal. In the subsequent step, small regions of the image that contain celestial objects are proposed using an FPN. With the feature maps corresponding to each proposed region, two parallel networks are used to locate the bounding box that encompasses the source object and provide its classification. To train and test the network, mock data was created using the SkyMaker software [67], joined by real images captured by a wide field small aperture telescope (WFSAT) and hand-annotated by undergraduate students. The ground truth labels consisted of the coordinates of each vertex of the bounding box surrounding an object, and magnitude and label information.

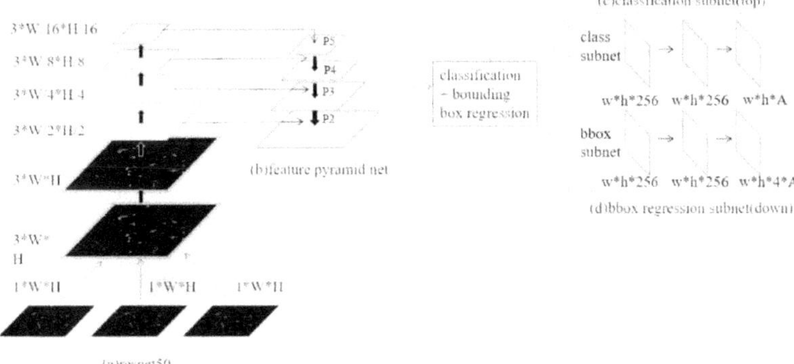

Figure 22 – Architecture of a Fast R-CNN based model for astronomical target detection and classification [66]

To evaluate the performance of the Fast R-CNN based solution, a comparison with the classical SExtractor software was performed. Figure 23 shows, on the left, the F1-score obtained for different values of an object's magnitude (with around 2 000 to 3 000 objects in each magnitude band), and on the right, the precision and recall attained. The authors note that although the classical technique and their proposal have similar results for sources with high signal-to-noise ratio, for lower ones their method is considerably superior. Despite this, they further add that when tested on simulated data, their approach was still not able to detect all the targets present, noting that considerable improvements are still possible for future developments.

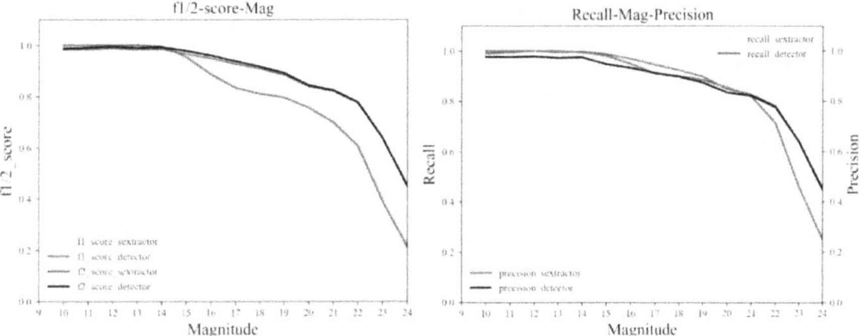

Figure 23 – Performance comparison between Fast R-CNN and classical (SExtractor) techniques for astronomical source detection [66]

3.3.2. ASTRO R-CNN

Similar to the method described in Chapter 3.3.1, Astro R-CNN [68] seeks to both detect and classify astronomical sources as being either galaxies or stars. However, instead of relying on the Fast R-CNN architecture, it uses the more modern Mask R-CNN. Its architecture, detailed in Figure 24 , relies on the ResNet-101 as the backbone, responsible for extracting relevant features from the source images, with initial layers recognising low-level features such as edges and corners, and the final layers recognising higher-level semantics such as galaxies or stars. FPN is similarly used to define hierarchically sized anchors used in multiscale object recognition. With the feature map selected, the region proposal network is tasked with the generation of two outcomes for each anchor: a class (positive or negative if an object is present or not) and a bounding box refinement (for the positive cases). For the final instance segmentation step (the masking), every pixel is assigned a class to mask the source objects and produce deblended cut-outs from full-scale images.

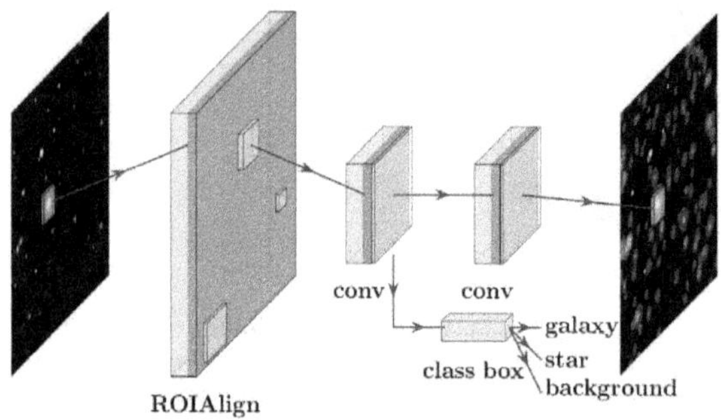

Figure 24 – Simplified overview of the architecture used by Astro R-CNN [68]

To overcome the problem of insufficient labelled information, synthetic data was generated using the PhoSim software (see Chapter 3.1) to create representations of both galaxies and stars. 512 x 512 pixel^2 images were created with stellar and galactical magnitude ranging between 12 and 23 and signal-to-noise ratio not lower than 2. Data augmentation was performed using: a) image and mask rotation by a multiple of 90 degrees; mirroring (left-right or up-down); b) blurring using a two-dimensional Gaussian kernel; c) adding or subtracting random values to each image channel. A total of 50 epochs were used in training, across two stages: the first only re-trained the head layers with a learning rate of 10^{-3}, and the second re-trained all layers with a learning rate that started at 10^{-4} and progressively decreased to 10^{-6}. 1 000 synthetic images (as illustrated in Figure 25) were used for training, 250 for validation and 50 for testing.

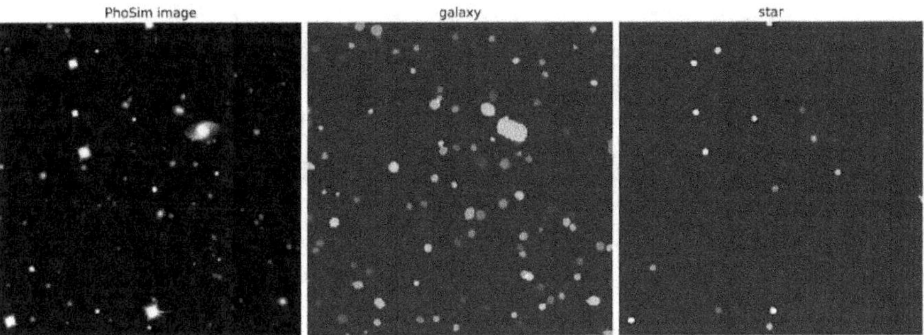

Figure 25 – Example of a synthetic image(left) generated for Astro R-CNN using the PhoSim package. The mask for galaxy classes (centre) and star classes (right) are also shown [68]

The results from the Astro R-CNN model were evaluated using the average precision (AP) score[2] metrics, which denotes the area under the precision-recall curve, averaged over each object at discrete recall levels. Table 4 summarises the scores obtained at each decision threshold and for different object sizes. Regarding the classification task, a total of 585 stars and 6 302 galaxies were correctly identified, 39 stars were erroneously predicted as galaxies and 78 galaxies as stars for an intersection-over-union (IOU) of 0.5 (or 50%), resulting in a precision of 88.2% and a recall of 93.8%. An example output of the Astro R-CNN is illustrated in Figure 26, where the green boxes represent stars and the blue boxes galaxies. Attention is drawn to the bottom right corner where the model displays good performance in crowded regions, successfully deblending the sources of various close-by objects.

Class	AP	AP_{50}	AP_{75}	AP_S	AP_M	AP_L
Star	48.6	86.8	32.3	86.8	-	-
Galaxy	49.6	83.9	42	75.3	7.6	0.3
Combined	49	84.2	39.8	76.4	7	0.3

Table 4 – Results obtained by Astro R-CNN on the test set. AP is the average score for IOU thresholds ranging from 0.5-0.95 with step 0.05. AP_{50} and AP_{75} are the AP score at thresholds 50% and 75%, respectively. AP_S, AP_M, AP_L are the AP scores of small-sized (area < 16^2 pixel^2), medium-sized (16^2 < area < 32^2 pixel^2) and large-sized (area > 32^2 pixel^2) objects, all at IOU bounding box threshold of 50%.

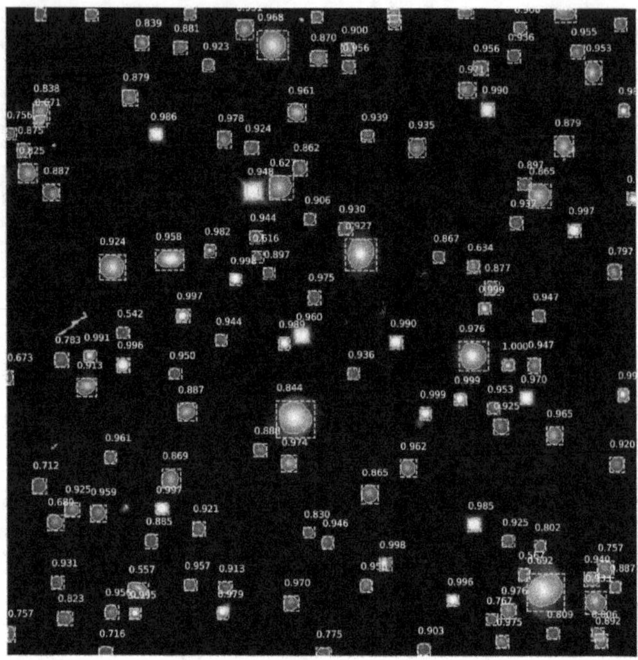

Figure 26 – Example output of the Astro R-CNN

3.3.3. APSCNET

The astronomy photometric source classification network (APSCnet) [69] is a source detection network based on the single-stage detector method YOLO v4. It not only distinguishes between stars and galaxies, as the previous methods, but is also capable of identifying quasars. Its architecture is highlighted in Figure 27.

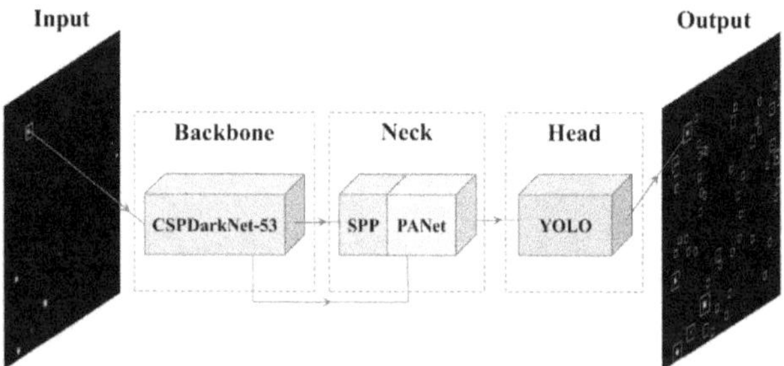

Figure 27 – Simplified overview of the architecture of APSCnet [69].

To generate image data a generative adversarial network (GAN) is used. Cropped sources from real image data are combined with synthetically generated backgrounds to create the final images. A generator network draws samples from random noise and learns to translate them to frames resembling real astronomical background, while attempting to fool a discriminator network whose job is to distinguish between real and generated background images. A total of 915 images are assembled for training, 183 for validation, and 124 real images are used in testing.

The source detection phase uses a CSPDarkNet-53 backbone for feature extraction. The feature pyramid stage ("Neck" in Figure 27) makes use of an SPP network (see Chapter 2.2.1.2) and path aggregation network [70]. Finally, the detection and classification layers are based on the YOLO architecture covered in Chapter 2.2.2.1.Results on the test set denote an AP score when the threshold is set to 0.5 (or 50%) of 88.0, surpassing the 84.2 obtained by [68]. Regarding the classification task, the results obtained by the ASPCnet are summarised in Table 5.

Table 5 - Summary of the classification results obtained by the ASPCnet on the test set [76]

		Prediction			
		Quasar	Star	Galaxy	Recall
	Quasar	1894	86	20	94.7%
Truth	**Star**	51	1947	2	97.4%
	Galaxy	15	1	1984	99.2%
	Precision	96.6%	95.7%	98.9%	

3.4. KEY REMARKS

Several advancements are currently taking place on the different phases of generation, modelling, and detection of stellar objects. Highly realistic methods for the creation of synthetic images may find a very important use in this type of studies, where they allow for deep learning models to be trained with known ground truths that are increasingly similar to real images of space. Likewise, a better representation of the PSF produced by stars may also contribute to improve the ability of novel learning-based approaches to measure the photometric and astrometric properties of these objects. Finally, several approaches for the application of deep learning to problems of astronomical object detection are being published over recent years. However, these are typically focused on tasks of object detection and classifications (i.e., distinguishing stars from galaxies and other astronomical objects), rather than in inferring their photometric properties.

This study therefore proposes to build on the problem of astronomical object detection, taking the current state-of-the-art approaches as a stepping-stone upon which to propose new methods and strategies to tackle a larger set of problems. Staring from the base idea of detecting an astronomical object in space, this study will expand on methods to understand how this detection holds for stars of different brightness and different crowding conditions. Additionally, new ideas will also be proposed to measure the centre of each star more accurately, as well as to calculate its flux.

4
METHODOLOGY

This study aims to develop an experimental study that can evaluate the performance of novel deep learning techniques for stellar detection and basic photometric analysis and compare it with the performance of traditional techniques based on algorithmic principles. Although the end-goal is to apply these methods to crowded stellar fields at the centre of the Milky Way galaxy, for this study, the simpler stellar cluster Collinder 140 will be used instead.

4.1. PROBLEM DEFINITION

Before any strategies or techniques are outlined, it is of critical importance to precisely define the problem. Be it through the use of methodologies such as CRISP-DM or more recent variations and extensions [71, 72] the phase of business understanding and translation of the business problem to the data mining goals are still of critical importance. For the scope of this study, the business objectives can be interpreted as the difficulties scientists face in their daily research and the data mining goals are described as the translation of these high-level objectives to the context of data science.

4.1.1. SCIENTIFIC PROBLEM

As outlined in Chapter 1, the key objective motivating scientists and astronomers working in this field is to develop a better understanding of the laws guiding the behaviour of black holes. Since these objects cannot be studied directly, as they do not emit any light, the best available option focuses on the study of the stars that orbit them. By understanding these interactions, scientists can better estimate some of the key characteristics of black holes, such as their mass, rotation, or spin.

For simplicity, this study will not focus directly on the detection of stars surrounding Sagittarius A*, but instead on a nearby open stellar cluster named Collinder 140. Albeit likely having a smaller density of stars than the region surrounding the centre of our galaxy [73], many of the same challenges concerning detection and deblending of point-sources are present in both scenarios, making Collinder 140 a stepping stone towards the final problem.

Two key strategies can be adopted to improve the ability to detect and more precisely study this region of space. The first, perhaps more expensive, consists of the developing and deploying better capturing equipment. New advancements in both ground and space-based telescopes, with higher image resolution and less susceptible to atmospheric perturbations, could have a very high impact on the scientists' ability to better detect and study a larger number of stars. Despite many new advancements in the field, these are still very expensive and highly technical tasks that typically require years, if not decades, of precise development and planning.

The second strategy, and the one explored in this study, consists of the development of new techniques capable of achieving better performance in the detection and characterization of the objects captured by the existing instrumentation. Although many physical and chemical properties could be inferred for these types of systems, the focus of the present study will consist exclusively of measuring the accuracy of the proposed method, determining the central coordinates of the existing stellar objects, and measuring the flux emitted by them. Images of this region of space are characterized by crowded stellar fields and are undisturbed by the presence of any other astronomical objects, such as galaxies, as illustrated in Figure 28.

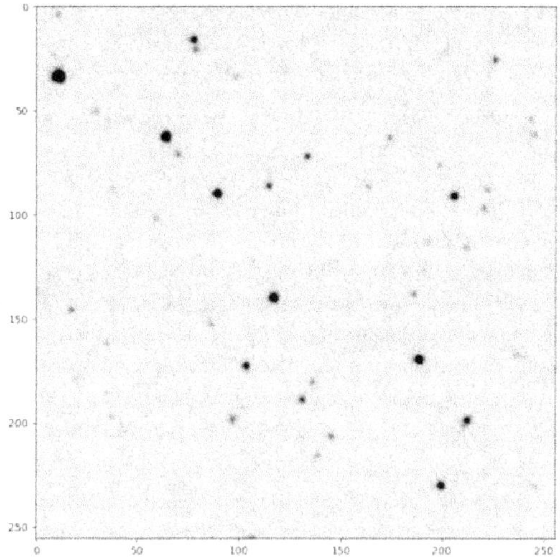

Figure 28 – Image of the Collinder 140 open stellar cluster captured by the Visible and Infrared Survey Telescope for Astronomy (VISTA) of the ESO's Paranal Observatory [74]. Shown in logarithmic scale and clipped at 10 standard deviations from the minimum electron count.

The scientific objectives for this study can therefore be described as follows:

- Detect individual stars present in stellar fields.

- For each detected star, find the coordinates of its centre of intensity as well as its flux.

- Compare the performance of the proposed methods with the classical techniques used by astronomers for photometry and stellar detection.

4.1.2. DATA SCIENCE STRATEGY

With the scientific objective clearly outlined, the definition of the data science strategy becomes the next critical task. Stellar fields that will be used as the main input for this study can originate from two distinct sources. One approach would be to collect real-world images captured by telescopes and use them as input to the machine learning model. The disadvantage of this approach is that images would be unlabelled or at best, labelled by human specialists or proven software. Solely using this approach would have the drawback that the model could only become as good as the weakest link in the chain, which in this case would be the lack of known ground truth samples. An alternative approach is to generate synthetic images that can closely approximate the characteristics of real stellar clusters. With that, the ground truth regarding the position and flux of the generated stars are fully known and can be used to train and better inform the model. Since the formation of images of this kind has been extensively studied (see Chapters 3.1 and 3.2) and equations that produce very close approximations to the real phenomena are known, this approach becomes a viable and desirable alternative to the first option. Of course, the drawback is now that the equations used for image generation are only approximations, and even if a model performs well on synthetic images, it is not guaranteed that performance will be kept in real ones.

The next step in outlining the data science strategy is to select a modelling approach. The state-of-the-art models described in Chapter 2.2 are designed to perform object detection and instance segmentation by not for regression on numerical properties of the detected objects. These models can be used for detecting an object and finding its centre of intensity, but they wouldn't be able to propose a value for the flux of the detected object, not without some alteration to their inner workings at least.

Two approaches are possible here. The first would be to divide the process in two steps. In the first step, the objects would be detected, and bounding boxes proposed for each. With that, a new model could be trained to solely perform regression on the extracted boxes to identify the flux of each star. This stepped approach is intuitive for humans, however, by breaking up the process into two stages, it runs the risk that during the optimization of the first stage (detection and extraction of bounding boxes), the model deviates from what would be considered the ideal path for the second stage (flux regression). This can happen since optimizing two loss functions separately might not lead to the same result as optimizing a single function that accounts for the full objective. For this reason, a second approach was chosen where the existing algorithms are modified to add an additional predictive head. That allows them to compute the regression to a given continuous value, alongside the already supported predictions. This has the advantage of maintaining the end-to-end approach with a single objective function to be optimized.

Finally, a baseline needs to be established to ascertain whether the proposed model can achieve any significant improvement compared to the state-of-the-art classical methods. Analytical approaches described in Chapter 2 will be used for comparison. This assessment will focus on three key factors: the number of true positives, true negatives, and false positives; the error in the predicted centre of intensity of the stars; the error in the predicted flux of the stars.

In summary, the following points contain the core skeleton of the proposed data science strategy and goals:

- Implement algorithms that generate representative synthetic images of the galactic centre.

- Implement traditional detection methods to serve as baseline for the study.

- Modify the architecture of existing object detection deep learning models to support the additional regression for the stellar flux.

- Modify the loss function of the modified deep learning model to suit the needs of this problem.

- Evaluate the performance of the new technique in comparison with the baseline models.

4.2. COLLINDER 140 DATASET

Collinder 140 (see Figure 28) is an open star cluster located in the constellation of Canis Major, consisting of a group of reasonably spaced stars of moderate brightness [75]. The Collinder 140 dataset [76] was captured by the Visible and Infrared Survey Telescope for Astronomy (VISTA) of the ESO's Paranal Observatory, located in Atacama region of Chile. The dataset produced comprised individual images captured by each of the 16 detectors, as well as a full image composed by stitching together the output of each detector.

Each individual image is computationally interpreted as a 2D pixel matrix (approximately 2000x2000 pixels in size) where each value corresponds to the electron count at each pixel. Although theoretically easier to interpret, the combined image of the 16 sources may have problems of stitching, i.e., when combining distinct regions of the sky, the space between them may need to be filled with a default value, or if they overlap, the intersections between them may also cause image defects. Therefore, this study will use the 16 individual images for the analysis where, due to computational limitations, each image will be further divided into a 256x256 pixel-sized grid.

The Collinder 140 dataset was also used as the basis from which the synthetic dataset generation method was constructed. Albeit with a set of assumptions and generalizations (detailed in Chapter 4.3), the key goal was to develop a generation process that accurately represented the distribution of stars, brightness and shapes that are found in the real world, without being too overly specific to the characteristics of this open cluster.

4.3. SYNTHETIC DATASET AND DATA GENERATION METHOD

Following on the strategy outlined by PhoSim [61], the data will be generated starting from a blank image by adding each phenomenon separately. In this way, each part of the astronomical process can be described individually, and the final image will contain the contribution of each phenomenon. The characteristics used to generate the synthetic images will be obtained from the real images of the Collinder 140 cluster. The key idea consists of describing how each process affects the electrons collected by a single pixel of the capturing equipment. Each pixel will start with a count of 0 electrons, and as each phenomenon is considered the count of the corresponding pixel is increased by the number of newly arrived photons.

4.3.1. SKY BACKGROUND

The sky background noise was estimated from the information available in the Collinder 140 dataset [74]. Initially, the median of the pixel values (corresponding to the electron count) for the full dataset was calculated alongside their standard deviation, resulting in the distribution shown in Figure 29.

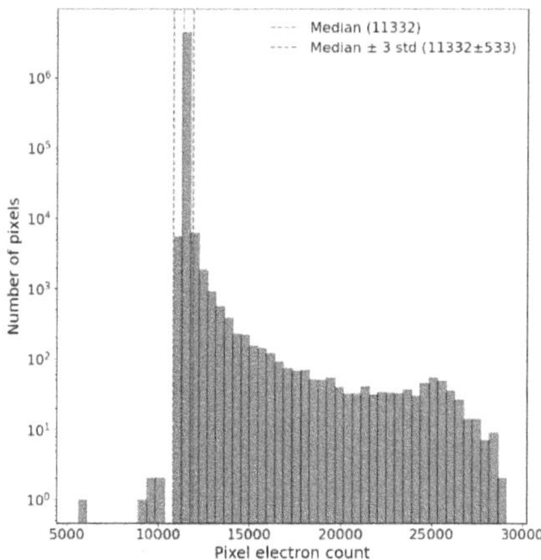

Figure 29 – Histogram of pixel values (pixel electron count) for the full Collinder 140 dataset.

The standard deviation of this estimate is largely influenced by the greater electron count of bright stars, leading to an unfair estimate of the true sky background noise. To correct for this, the distribution in Figure 29 was filtered to only include values within 3 standard deviations of the median. This resulted in the new distribution shown in Figure 30, where a much better approximation can be seen, resembling a Gaussian distribution. From here, the mean of the sky background noise can be easily estimated as 11332 electrons and its standard deviation as 33 electrons.

For the generation of the synthetic images, the value of the sky background for each pixel was taken as a value sampled from a Gaussian distribution, with a mean of 11332 electrons and a standard deviation of 33 electrons.

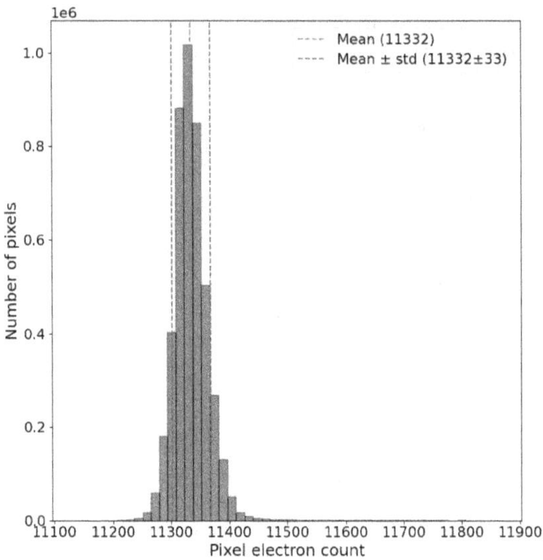

Figure 30 – Estimate of the background noise mean and standard deviation for the Collinder 140 dataset.

4.3.2. DISTRIBUTION OF STELLAR OBJECTS PER MAGNITUDE BAND

As part of the Collinder 140 dataset [74], information regarding the detection of stars was also made available. These objects were detected using the classical photometry techniques discussed in Chapter 2.1. Although they have been successful for many decades, their detection capabilities are largely restricted by the brightness of the objects above the background (signal-to-noise ratio) and the degree to which sources overlap. The magnitude of the detected objects is a proxy for the flux used by astronomers. It was estimated using equation (4.1):

$$m_k = -2.5 \log_{10}\left(\frac{F}{F_{ref}}\right)$$

(4.1)

where m_k is the magnitude of the measured star, F the flux of the observed star, and F_{ref} the flux of a reference star observed (typically Vega) with the same instrument and corrected for any perturbance conditions. Although the value of F_{ref} was not directly provided in the catalogue, it was estimated to be around 1.41×10^{10} based on the flux and magnitude of the provided observations. With this, the magnitude of each star could be estimated, resulting in the distribution shown in blue in Figure 31. It can be easily noted that for magnitudes higher than 17 (meaning fainter stars), the photometric software starts having difficulties in detecting the objects, while between magnitudes of 11 and 17 the number of stars increases linearly (in the logarithmic scale) with the increase in magnitude. To approximate the expected true count of stars, a linear approximation (on the logarithmic scale) was performed to the photometric detections between magnitudes 10 and 17. The counts of stars for all magnitudes between 10 and 20 was then interpolated, resulting in the distribution shown in magenta in Figure 31.

Each unit increase in magnitude represents a power-law decrease in the measured flux of a given star, and by consequence, also in its measured amplitude. This means that fainter stars are considerably dimmer than bright ones and much more difficult to detect. As the number of stars also increases in power-law with the decrease in brightness, it can be expected that the environment in which the deep learning algorithms will be trained will be considerably more challenging than then if a distribution closely matching the photometric results was assumed.

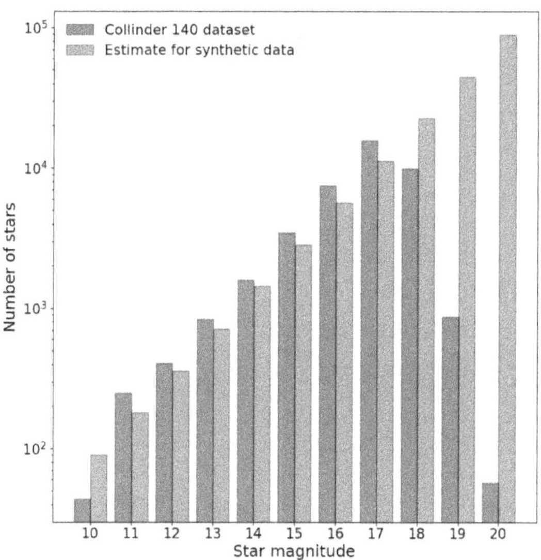

Figure 31 – Distribution of the number of stars per magnitude band from the Collinder 140 photometry catalogue and estimated approximation to be used for the synthetic dataset. Number of starts shown in logarithmic scale.

4.3.3. DISTRIBUTION OF STELLAR OBJECTS PER IMAGE

From the available pixel dimensions of each image in the Collinder 140 dataset and the total number of detected stars, the average density of stars was estimated to be 37.5 per square region of 256x256 pixels. With the estimated true count defined in Chapter 4.3.2, the same logic would yield a density of 176 stars per region of 256x256 pixels. To allow for some variability between different regions of space, the synthetic images were generated with the number of stars for each being randomly taken from a uniform distribution between the values of 35 and 180 stars per square region of 256x256 pixels.

4.3.4. ESTIMATION OF STAR SHAPE

The photometric information provided in the Collinder 140 catalogue contains data regarding the shape of each detected star, namely the standard deviation of its Gaussian-like form. Grouping this information by magnitude, it is possible to note that the radius of larger stars is greater than their dimmer counterparts. Figure 32 shows, in blue, the different estimates for the Gaussian sigma for the different magnitude groups. The sudden increase in sigma for the stars of magnitude 10 and 11 is likely due to the pixels reaching their maximum well depth and the image saturating locally. Still, in Figure 32, the estimates for both the mean and standard deviation of the sigma values used for generating synthetic images are shown in magenta. These were computed following the same procedure described above, where the values for magnitudes down to 12 were approximated by a linear function, which was later used to calculate the properties for each magnitude.

In the image generation process, the sigma of each star is selected from a Gaussian distribution with the mean taken as the value of the corresponding magnitude and having the respective standard deviation. This ensures some variability in the shape and form of the resulting stars and, as the standard deviation of each magnitude overlaps with that of its neighbours, ensures that no clear jump exists between magnitudes as well as that the model cannot rely solely on this information to extrapolate the flux of a given star.

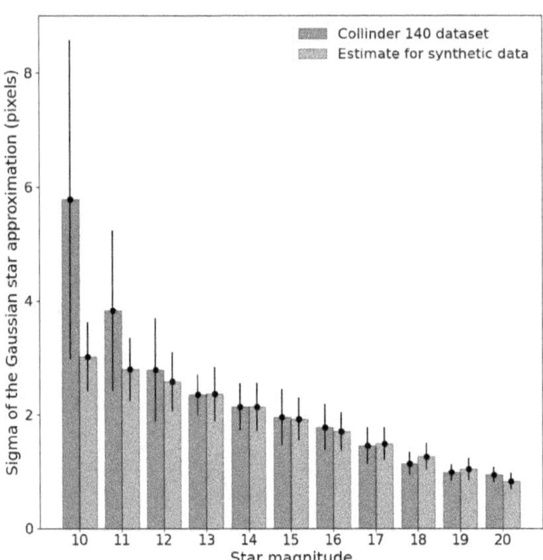

Figure 32 – Distribution of the sigma mean and standard deviation for the Collinder 140 dataset. Data from the original photometry is shown in blue and the estimations to be used for image generation in magenta.

4.3.5. PERTURBATIONS IN THE STAR POSITION

Due to the vast distances between cosmological objects, in one night of observations of a given region of space, all the objects measured from our observer position will seem as if they have not moved at all. However, slight variations in their positions can still be observed even between images captured only a few minutes our hours apart. This is due to atmospheric effects and other perturbations that can affect each image differently and result in a star appearing in a slightly different location from frame to frame.

To account for this effect, and since the images generated will have units of photo counts, with no representation of colour channels, the notion of channels will instead be used to produce slight variations of the same image where the position of each object is slightly altered. To achieve it, a value sampled from a uniform distribution ranging from -1 pixel to $+1$ pixel was added to each coordinate of each star. The final image will contain 4 channels in which each star appears in a slightly different position, simulating the intended effect. Regarding the ground truth star coordinates, the original coordinates before this effect is applied are saved and will be later used for training the model as they reflect the true position of the object, even though they may not be present in any of the four channels.

4.3.6. POINT SPREAD FUNCTION AND GENERATION OF SYNTHETIC STARS

As detailed in Chapter 2, the shape of the star measured by the instrument is dictated by the point spread function that characterizes it. Due to its simplicity and good representation of many real-world scenarios, the Gaussian function will be the formulation used for the generation of the star PSFs. The main steps taken during this process can be summarized as follows:

1. Select a random number from a uniform distribution between the minimum and the maximum number of stars in a 256x256 pixel image (see Chapter 4.3.3)

2. Generate two tensors, one for each star coordinate on the image plane, sampled from a uniform distribution between the minimum and maximum pixel values.

3. Create 4 channels for each of the above tensors, varying the coordinates of each star slightly, to represent similar images of the same region of space where the star position change slightly due to atmospheric effects and other perturbations (see Chapter 4.3.5)

4. Take a sample, with size equalling the number of stars, from the magnitude distribution described in Chapter 4.3.2 (note that this distribution was actually approximated by a continuous function), where the probability of each magnitude value being sampled equals the count of stars with that magnitude over the total count of all the stars (integral or area under the curve).

5. Calculate the expected flux for each of the sampled magnitudes by solving equation (4.1) to extract the flux F.

6. Take a sample from a Gaussian distribution with the mean and standard deviation corresponding to the respective magnitude of the star (see Chapter 4.3.4) to find the value of sigma for each star.

7. Calculate the amplitude of each star by using equation (4.2) where A is the amplitude, F is the flux estimated in step 5, and sigma is the value estimated in step 6.

$$A = \frac{F}{2\pi\sigma^2} \tag{4.2}$$

Having performed all the previous steps, the final shape of the star and the value of each pixel in the image can be calculated using a Gaussian PSF, as detailed in equation (4.3):

$$f(x, y) = A e^{\left(\frac{-(x-x_0)^2-(y-y_0)^2}{2\sigma^2}\right)} \tag{4.3}$$

where x and y are coordinates of a given pixel, x_0 and y_0 are the coordinates at the centre of the function and $f(x, y)$ would yield the value of a pixel at a given x, y coordinate. Lastly, the final value of each pixel in the PSF is taken from a Poisson distribution, where the rate parameter is the original pixel value, as calculated by equation (4.3).

The image for each star is generated individually from a 256x256 matrix filled with zeros by adding the result of equation (4.3) plus the Poisson sampling for each pixel. Subsequently, the matrices for all stars are added together meaning that for the pixels where their shapes overlap, the final pixel value will include each stars' contribution, following on the principle of superposition.

4.3.7. TRAINING, VALIDATION AND TEST DATA

The use of synthetically generated images allows for new, never seen before images, to be shown to the model each time a training sample is required. This will ensure the model is constantly learning from previously unseen data and prevents the risk of overfitting to a finite set of data. During the experiment phases where the model is being trained and the best hyperparameters are being searched, a validation set of 128 unseen synthetic images will be used to evaluate the final performance of each model at the end of training. This synthetic dataset will be the same for all the models, so a fair comparison between them can be established. Once the final hyperparameters have been defined and the model selected, the evaluation will be performed in a yet unseen test dataset composed of 256 synthetically generated images. Figure 33 shows a schematic representation of the training, validation, and test flows, alongside the image generation process.

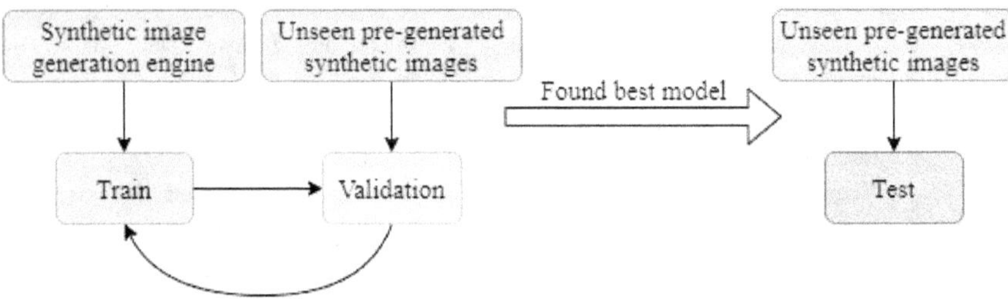

Figure 33 – Schematic representation of the training, validation, and test flows together with the synthetic image generation process used in the development of the deep learning models.

4.4. MODEL DEVELOPMENT

From the literature review conducted in Chapter 2, the region proposal-based frameworks currently display a slightly superior performance compared to the regression/classification-based methods on generic object detection problems. Furthermore, recent applications of deep learning object detection algorithms to problems of detection of astronomical sources [66, 68] have been developed with the use of region-based convolutional neural networks such as Fast R-CNN and Mask R-CNN. The most recent, and arguably better performant model of this family is the Mask R-CNN. As discussed in Chapter 2.2.1.6, the improvement it promoted over its predecessor (the Faster R-CNN) and its major selling point is the prediction of pixel masks around the detected objects instead of a bounding box.

Due to the nature of the problem being studied, covered in detail in Chapter 4.1, no particular benefit has been identified for predicting the pixel mask of a star instead of its bounding box. In fact, even the prediction of a bounding box is simply a mechanism to extract the centre coordinates of the star. From an astronomical perspective, stars would behave as pure point-sources if not for the various atmospheric effects that affect their shape and give them the PSF form discussed. Considering all these factors, and since the Mask R-CNN would bring no clear advantage for this problem (in fact it could be even more complex to determine the centre or irregular pixel shapes), the decision was made to use the Faster R-CNN architecture for this problem.

4.4.1. MODEL ARCHITECTURE

The architecture used for the Faster R-CNN framework is largely based on the existing implementation available in *PyTorch* [77], with some important modifications to adapt it to the current problem. The network consists of an initial fully convolutional layer together with a Feature Pyramid Network, responsible for extracting the relevant features from the original image and producing feature maps at different scales or levels, respectively. From these feature maps, a Region Proposal Network can generate a set of likely regions where objects are expected to be located. The Region of Interest Align phase is responsible for mapping the proposals made at different scales back to the desired location in the original image. Finally, the predictive head is responsible for processing all the extracted features and inferring the final output predictions. For the case of the present study, this would be the classification (if the object is a star or not), bounding box regression (the coordinates of the bounding box whose centre corresponds to the centre of intensity of the star), and flux regression (the value of the flux of a given star). A schematic representation of the implemented architecture is illustrated in Figure 34.

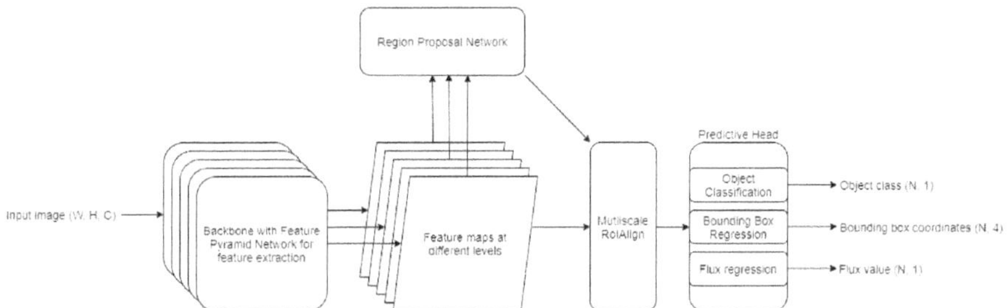

Figure 34 – Architecture of the implemented Faster R-CNN network. W, H and C represent the width, height and channels of the input image, respectively, while N represents the number of objects present.

4.4.2. PREDICTIVE HEAD

As shown in Figure 34, the predictive head of the Faster R-CNN network had to be altered to account for the requirement of the prediction of the star flux. The input for the predictive head is a one-dimensional tensor of size 1000 from which the desired outputs (score, bounding box coordinates and flux) must be derived.

The implementation of the original predictive head relies on the confidence score (object classification) the model attaches to each prediction to filter the remaining values of the bounding box coordinates appropriately. This behaviour was maintained for the implementation in the present work where the score was also used to filter the proposed values for the flux regression ensuring all the outputs are of the same size. Therefore, a summary of the architecture used for the predictive head is shown in Table 6.

Table 6 – Summary of the adaptations made to the predictive head to adapt it for this task

Name	Activation function	Input shape	Output shape	Modifications
Score	Linear	(N, 1000)	(N, 1)	None
Bounding box	Linear	(N, 1000)	(N, 4)	None
Flux	Linear	(N, 1000)	(N, 1)	Added

Variations of this architecture for the predictive head will be evaluated in section 4.6, where instead of going directly from the input layer of size 1000 to the output layer of size 1, intermediate layers are added. Two additional variations are proposed. On the first, a single intermediate layer with 512 neurons is introduced between the input and output. In the second, two intermediate layers are added between the input and the output, with the first having 512 neurons, and the second 128. The activation function for these intermediate layers will be the Rectified Linear Unit (ReLU).

4.4.3. LOSS FUNCTION

Since a new prediction was added to the model, the loss function that will be used to guide the learning process also needs to be altered to account for this change. The original implementation [77] makes use of the cross-entropy for the loss of the classification head and the smooth L1 loss for the regression of the bounding box coordinates. Given that the newly introduced head will also implement a regression, the commonly used approaches include the L1 loss (or mean absolute error), the L2 loss (or mean squared error) and a combination of the two, the smooth L1 loss, shown in equation (4.4):

$$smooth_{L_1}(x) = \begin{cases} 0.5x^2 & if \ |x| < 1 \\ |x| - 0.5 & otherwise \end{cases} \tag{4.4}$$

which is less sensitive to outliers than the mean squared error and, in some cases, can prevent the problem of exploding gradients [41]. As detailed in Chapter 4.3, since the star fluxes grow exponentially with decreasing magnitude and for some of the brightest stars, the image may reach its saturation limit, the smooth L1 loss is also chosen for the flux regression.

4.5. TRAINING AND DEVELOPMENT STRATEGY

A strategy will be required to guide the development of all the code and logic required to conduct this study, both in terms of the models utilized, as well as the images constructed. This continuous development cycle will aim to identify key hyperparameters, image processing options, and other design approaches that will later be broadened and expanded in the experiments (see Chapter 4.6) and evaluation (see Chapter 4.7) sections.

4.5.1. DATASET AND INPUT IMAGES

During the training phase, only the synthetic images described in Chapter 4.3 will be used. New images will constantly be created in real-time as the learning progresses, for both training and validation, meaning that the model will always be learning from previously unseen data. The image size used for all cases will be 256x256 pixels. Although the images from the Collinder 140 dataset are typically around 2000x2000 pixels, limitations on the software available (particularly GPU RAM) make larger sizes unviable. Table 7 summarizes the key decisions used during training and development.

Table 7 – Summary of dataset and input images characteristics used during the development phase.

Process	Method	Chapter
Image size	256x256 pixels	
Sky background	Pixel values randomly taken from a Gaussian distribution with mean of 11332 and standard deviation of 37 electrons	4.3.1
Number of stars per image	Taken from a uniform distribution between 35 and 180 stars per 256x256 pixel region	4.3.3
Star amplitude	Taken as from an approximation of the estimated distribution of stars per magnitude	4.3.2, 4.3.6
Star sigma	Taken from a Gaussian distribution with mean as standard deviation as per Figure 32 for the corresponding magnitude	4.3.4
Perturbations to the star position	A change from the original position taken from a uniform distribution between -1 and +1 pixel	4.3.5

4.5.2. MODEL

With the Faster R-CNN selected as the main architecture to be used, several decisions still had to be made regarding all the components used to construct it. Originally developed with a VGG backbone, modern R-CNN implementations typically use the Resnet for feature extraction instead [42, 78]. The residual connections it implements allow for deeper networks to be trained while minimizing the problems of vanishing gradients and degradation. Due to its smaller size and RAM requirements, the ResNet50 architecture will be used during the development. The remaining characteristics, including the feature pyramid network, region proposal network, region of interest alignment, predictive head and loss function will be used as detailed in Chapter 4.4.

4.5.3. TRAINING PARAMETERS

The training during the development phase will consist of a maximum of 50 epochs with a batch size of 256 images, resulting in a total of 12800 images at the end of the training process. The mini-batch size used will vary between 4 and 32, always consisting of multiples of 2. At the end of each epoch, the model will be evaluated on the validation set, consisting of a batch of 32 previously unseen images, with a mini-batch size of 4. Due to the computational complexity of some of the metrics that will be evaluated (see Chapter 4.7), it was noted that larger validation batch sizes would considerably impact the speed of development.

Several tries were made to find a suitable optimizer for this stage. Initial attempts with the Stochastic Gradient Descent (SGD) approach resulted in poor performance, leading to a more challenging development. Later investigations yielded more promising results with the Adam [79] optimizer and even slightly superior with the AdamW [80], using a learning rate of 1×10^{-4}. The development cycle will therefore use these two hyperparameters, with a constant learning rate. The gradients will be calculated at the end of each mini-batch and the parameters updated accordingly via backpropagation.

Regarding the loss function and weights of each loss, the procedure highlighted in 4.4.3 will be followed, with each of the three parts that constitute the total loss having the same weight. The reduction or aggregation method selected will be the sum.

4.5.4. CHALLENGES ENCOUNTERED DURING DEVELOPMENT

One of the major problems encountered during the development of the framework used to conduct this study was the large differences between the pixel values of the brightest stars, and those of the faintest ones. Not only are the brightest stars orders of magnitude brighter than their dimmer counterparts, but they are also orders of magnitude rarer. This becomes particularly problematic for the model, since the methods used to identify and locate the brightest objects in an image do not yield good results for faint objects. The result is a very poor performance of the model, where only a handful of stars, on the lower regions of the magnitude distribution are found. An example of one unaltered image initially fed to the model, is shown in the top image of Figure 35. Although only a handful of stars can be clearly visible to the naked eye, this image contains a total of 106.

This high discrepancy between a very small number of extremely bright stars and a large number of very faint ones was thought to be causing increasing difficulties in the learning process. An attempted way to solve this problem was to represent the pixel values within an image on a logarithmic scale. Although this would seem to address the issue and does improve human visibility regarding the number of stars in an image, it still did not produce the expected results on model performance. It is difficult to know exactly why the detection performance did not improve with this technique, but likely factors may be that, as the differences between pixel values get compressed, it may become more difficult to distinguish between a star and sky background noise.

The next method attempted was to perform clipping of the pixel values within a certain standard deviation of the calculated median value of the sky background. An approach very similar to that described in chapter 4.3.1 was followed to estimate the median background noise and its standard deviation. This was performed twice for different bounds of standard deviation in relation to the median, namely from -2 to +2 standard deviations of the median for the image in the bottom-left image of Figure 35 and between +2 and the maximum pixel value for the bottom-right image. With this approach, the space between different values is maintained constant, with the obvious exception of the clipped values. This improves the visibility into several fainter stars, albeit still leaving many invisible to the human eye, with only around half of the total 106 being clearly visible. With this technique, the performance of the model increased substantially, as it was now able to locate the stars more easily. Evidently, the drawback of this approach is that the clipped stars no longer present a shape that can be approximated by a Gaussian curve, and their peak have been cut to form a large plateau around the clipping value. Although this shouldn't be a problem for the object classification and regression of the bounding box coordinates, it could potentially affect the predicted flux values. For this reason, the first 4 channels of the input were kept with the original unaltered image, while the next two sets of 4 channels contained a version of each of the clipped variants.

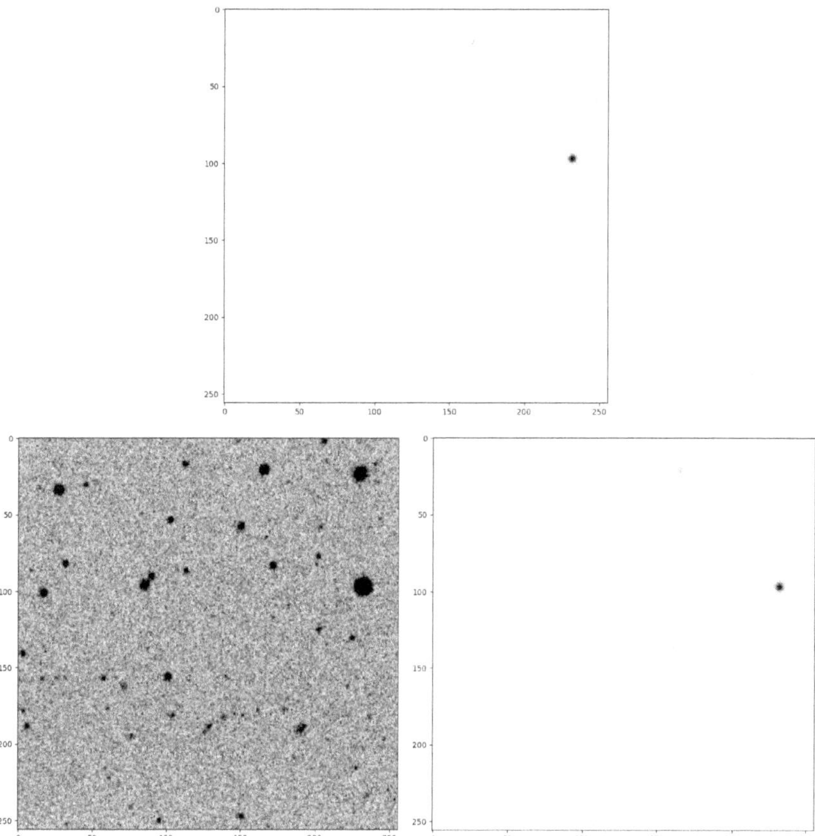

Figure 35 – Example of an unaltered synthetically generated image (top row). The same image is shown clipped between -2 and +2 standard deviations from the background noise on the bottom-left, and between +2 standard deviations to the maximum value on the bottom-right.

4.6. EXPERIMENTS

Once the development of the programming and logical phases of the study are implemented, the next step consists in running a series of experiments to identify the most suitable set of hyperparameters to be used. For the first phase, a broader spectrum of choices will be available, and several models will be trained for 50 epochs with batch size of 256 to measure their performance. On the second and final phase, the top performing models will be selected and trained for 200 epochs with 512 batch size using the best set of parameters previously found. Resulting that in phase 1 the models are train on a total of 12800 images, while on phase 2 this increases to 102400 images. Only synthetic images will be used for either phase. Table 8 summarizes all the hyperparameters in the search space. Due to the large number of possible hyperparameter combination, randomized search is used instead of exhaustive grid search.

Table 8 – Search space of hyperparameters used for the first experiment phase.

Hyperparameter	Type of distribution	Range of possible values
Backbone network	Discrete uniform	Resnet50, Resnet152, MobilenetV2, EfficientNetB4, Densenet161
Predictor architecture	Discrete uniform	Input-Output, Input-512-Output, Input-512-128-Output
Class loss weight	Continuous uniform	Minimum: 1 Maximum: 2
Bounding box regression loss weight	Continuous uniform	Minimum: 1 Maximum: 2
Flux regression loss weight	Continuous uniform	Minimum: 1 Maximum: 2
Mini-batch size	Discrete uniform	8, 16, 32
Optimizer	Discrete uniform	AdamW, NAdam, RAdam, Adamax
Learning rate	Continuous uniform	Minimum: 1×10^{-6} Maximum: 5×10^{-5}
Weight decay	Continuous uniform	Minimum: 1×10^{-5} Maximum: 1×10^{-4}

4.7. BASELINE MODELS

Two approaches will be used to establish a baseline comparison that can ascertain whether the deep learning model show a performance improvement over the traditional methods. In the first, the DAOPHOT algorithm discussed in Chapter 2.1.1 will be implemented and used for prediction in both the synthetic and the Collinder 140 dataset. This is intended to give a perspective of the conventional method on a dataset where the absolute truth is known and better understand its limitations and strengths. Additionally, and exclusively for the Collinder 140 dataset, the professional photometric catalogue provided will also be used.

4.7.1. DAOPHOT

This algorithm typically requires some manual bespoke work to find clear and well separated sources within an image that will be used to approximate the shape of the point spread function to use for all stars. To achieve a fair comparison between processes, this initial step was automated and can be described as follows:

1. Locate the brightest star in an image that is at least 10 pixels away from the edge (this can be done by finding the maximum pixel value)

2. Perform a horizontal cut centred at the brightest stars and extract five-pixel values either side of the centre.

3. Divide the maximum pixel value by 2

4. Calculate where the straight line defined in step 3 intercepts the curve of the star profile extracted in step 2.

5. Take the length between the two interception points as the FWHM of the PSF.

6. Convert the FWHM to the sigma of the PSF.

Even with the approximation of the sigma of the PSF in the image, there are a few extra parameters that still need to be estimated before the algorithm can be applied. To avoid a manual analysis of the images and perform a fair comparison between two automated processes, the remaining parameters of DAOPHOT will be trained, similarly to the deep learning framework. To achieve this, an exhaustive grid search will be performed across the parameters detailed in Table 9 and the best resulting model selected.

Table 9 – Summary of the hyperparameters of DAOPHOT learning through model training with grid search

DAOPHOT parameter	Type of distribution	Range of possible values
Threshold factor	Discrete uniform	3, 4, 5
Criteria of separation	Discrete uniform	0.5, 1, 2
Fit shape	Discrete uniform	11, 13

4.7.2. COLLINDER 140 CATALOGUE

The catalogue for the Collinder 140 dataset, provided alongside the images, consists of a photometric survey conducted by the researcher group at VISTA Hemisphere Survey project utilizing the Cambridge Astronomical Survey Unit (CASU) software [76]. Although not previously mentioned, this data processing and photometry package has a specific release for the VISTA project and is based on the same principles of the analytical techniques discussed in Chapter 2.1. The information available contained the pixel coordinates of the centre of each detected star, the measured flux at different aperture thresholds, and an indication of the magnitude of the detected stars.

4.8. EVALUATION

As discussed in Chapter 4.1, there are three key parts that ought to be evaluated in this study, these are the correct detection of an object as a star, the difference between the predicted centre coordinates of this object and its true value, and the difference between the predicted flux and its ground-truth. Several metrics will be employed to evaluate each of the different performance indicators.

4.8.1. PRECISION

The precision of a machine learning model is helpful in identifying the proportion of samples correctly labelled among those the model predicted as being of the positive class. It can be calculated from equation (4.5):

$$Precision = \frac{TP}{TP + FP} \tag{4.5}$$

where TP are the true positives and FP the false positives. It can also be understood as answering the question: "Of the total number of positive predictions made by the model, how many were correct?". Although this can be a useful metric, it fails to account for the objects that the model did not identify.

4.8.2. RECALL

The recall of a machine learning model helps addressing the previous shortcomings and can be used to identify how many of the total number of positive samples were correctly identified by the model. It can be expressed with the use of equation (4.6):

$$Recall = \frac{TP}{TP + FN} \tag{4.6}$$

where FN represents the false negatives. It can also be interpreted as answering the question "Of the total number of positive samples, how many were correctly identified by the model?".

4.8.3. F1-SCORE

As both precision and recall attempt to measure slightly different aspects of a model's performance, an approach to combine the two is provided by the F1-score. It is defined as their harmonic mean and can be calculated according to equation (4.7):

$$F1score = \frac{2 \times Precision \times Recall}{Precision + Recall} \tag{4.7}$$

For problems where it is important that the model can identify the maximum number of true samples while also minimizing its erroneous predictions, it may be considered as a more suitable metric than precision and recall individually. These three metrics will be mostly used to evaluate the model's ability to correctly identify objects in an image, i.e., for the classification task.

4.8.4. MEAN AVERAGE PRECISION

One of the most widely used metrics in problems of object detection is the mean average precision (mAP). When the goal is to localize an object in an image, simply stating that the object is present is no longer sufficient, a criterion must also be defined by which it can be ascertained whether the object was correctly located. The mAP starts with a calculation of the intersection between each proposed bounding box and the corresponding ground truth box, divided by the union of their areas. After defining a certain threshold between 0 and 1, proposals with an intersection over union (IoU) above the threshold are considered as correct detections and those which remain below it are labelled as false positives. The next stage consists in calculating the precision and recall for all objects in all images at the selected threshold. From there, the precision-recall curve can be estimated, and the average precision computed as the area under this calculated curve. Finally, the mean average precision is computed by averaging the average precision for each of the distinct object present in the images. Since in this study we will only be considering stars, both average precision (AP) and mean average precision (mAP) will result on the same value. This metric is also tightly connected to the classification and localization tasks and will be mostly used to assess the performance of different models and guide the development process.

4.8.5. ROOT MEAN SQUARED ERROR

To measure the ability of the model to accurately predict both the position of the centre of a star as well as its flux, a regression metric is necessary. The root mean squared error (RMSE) is one of the most widely used methods in these scenarios and starts by calculating the square of the error (or difference) between the predicted value and the ground truth. Since the errors are squared before their mean is computed, higher discrepancies between the predictions and targets will have larger impact on the result. Lastly, the square root is calculated to arrive at the final value for the RSME. Its formula is detailed in equation (4.8), where $\hat{y}_i$ is the predicted value for element i, and y_i is the ground truth.

$$RMSE = \sqrt{\sum_{i=1}^{n} \frac{(\hat{y}_i - y_i)^2}{n}} \tag{4.8}$$

4.9. KEY REMARKS

The problem definition starts by identifying the main objectives from a scientific perspective to then translate them to actionable tasks in the context of data science and computer vision. This is an important phase to ensure that the two scientific areas are aligned, as well as to guarantee that the outputs produced address all the necessary requirements for the proposed objectives.

The key idea of the synthetic dataset is to provide for an environment where the ground truth is known, so that the model can be trained using a supervised learning approach, as well as to create a train setting as independent of the biases of a single stellar cluster as possible. It builds on some of the ideas discussed in previous works for the generation of artificial images, albeit a simpler approach is used to ensure a fast development cycle with quick response to changes.

A conventional approach is proposed for training the deep learning model. In an initial phase, a larger number of architectures are tested with a broad spectrum of possible hyperparameters, from which the best performing approach are selected for a second a second experiment where the focus lies more on hyperparameter fine tuning rather than discovery. The baseline models are also described in detail as they will serve as a reference by which to compare the performance of the deep learning approach.

Finally, the main evaluation metrics are thoroughly detailed, as they will be of critical importance for the result analysis and for comparing the different methods testes. Metrics are proposed for all the three tasks that will be studied, namely, detection, localization, and flux prediction.

5
RESULTS

The various experiments performed are designed to guide the development and selection process in the initial stages and, as the most promising hyperparameters start emerging, to provide a comparison between the traditional methos (in this case DAOPHOT) and the deep learning approaches.

5.1. FIRST EXPERIMENT PHASE

Conducted exclusively for the deep learning models, the first experiment phase seeks to find the most promising combination of hyperparameters to be later explored on the second phase, as detailed in Table 8. Since during the development it was found that the backbone was one of the critical hyperparameters for the success of the model, a total of 10 experiments was run for each backbone, allowing the remaining hyperparameters to be randomly selected between the defined intervals. Table 10 fully details the results of this experiment where only the 3 best performing models for each backbone are shown, including both the hyperparameters used and the evaluation metrics obtained. To facilitate the comparison between the overall performance of the different backbones, Table 12 averages the obtained metrics for each. In both tables the results are ranked by mAP50 (mean average precision for an IoU threshold of 0.5 – see Chapter 4.8.4), in descending order.

By analysing Table 12, it is possible to note that the best performing backbone regarding mAP50 is the ResNet 50, closely followed by the ResNet 152. The inclusion of two backbone architectures of the same kind (their key difference lies in the depth of the extracted features) was intended to evaluate whether sufficient benefit was gained by including a deeper network (ResNet 152). Although the differences are not critical, the shallower architecture seems to have a small edge on the mAP50. The remaining architectures show a clearly worse performance on this metric.

Looking at the following three metrics (precision, recall and F1-score), the ResNet architectures continue showing a considerable superior performance than their counterparts, with the precision being slightly higher for the ResNet 50, but recall and F1-score being better for the ResNet 152 architecture. Finally, when analysing the RMSE of the flux prediction, the same picture emerges, with the two ResNet architectures performing considerably better than the remaining, with a slight edge again for the ResNet 50.

This page is intentionally left blank.

Table 10 – Summary of the hyperparameters and evaluation metrics of the three best performing deep learning models for each backbone in experiment phase 1.

Model name	Backbone network	Predictor architecture	Class loss weight	Bounding box regression loss weight	Flux regression loss weight	Mini-batch size	Optimizer	Learning rate	Weight decay	mAP50	Precision	Recall	F1-score	Flux RMSE (electrons)
0dbee24e	ResNet 50	B	1.43	1.30	1.77	8	NAdam	9.20×10^{-5}	7.30×10^{-4}	0.561	0.890	0.568	0.694	37 676
97d8635c	ResNet 152	B	1.05	1.99	1.31	8	NAdam	9.98×10^{-5}	1.73×10^{-4}	0.518	0.948	0.522	0.673	43 735
e20726a6	ResNet 50	A	1.27	1.62	1.58	16	AdamW	8.15×10^{-5}	3.25×10^{-4}	0.498	0.915	0.502	0.649	53 684
cddbdfce	ResNet 152	B	1.66	1.47	1.61	8	NAdam	8.24×10^{-5}	2.28×10^{-4}	0.474	0.968	0.476	0.638	55 928
00fbb505	ResNet 50	B	2.00	1.84	1.73	8	NAdam	3.68×10^{-5}	2.80×10^{-4}	0.462	0.906	0.469	0.618	54 585
b352f722	ResNet 152	B	1.19	1.73	1.42	8	Adamax	9.37×10^{-5}	5.79×10^{-4}	0.423	0.745	0.450	0.561	54 717
ea95f70c	DenseNet 161	C	1.64	1.89	1.25	16	NAdam	8.27×10^{-5}	5.76×10^{-4}	0.218	0.661	0.260	0.373	68 982
e320b84a	MobileNet V2	C	1.84	1.89	1.39	8	NAdam	9.98×10^{-5}	9.25×10^{-4}	0.062	0.454	0.117	0.186	78 914
7dad28fe	MobileNet V2	A	1.31	1.91	1.82	8	AdamW	8.18×10^{-5}	6.82×10^{-4}	0.048	0.454	0.091	0.152	87 464
6650c400	DenseNet 161	C	1.39	1.53	1.43	8	NAdam	2.43×10^{-5}	3.30×10^{-4}	0.030	0.391	0.071	0.120	108 335
0c1f039d	DenseNet 161	A	1.69	1.40	1.48	16	NAdam	3.56×10^{-5}	4.78×10^{-4}	0.020	0.404	0.050	0.089	128 453
26679b4c	EfficientNet B4	C	1.25	1.29	1.02	8	AdamW	6.53×10^{-5}	8.21×10^{-4}	0.019	0.316	0.044	0.078	86 166
74750d2f	EfficientNet B4	A	1.34	1.70	1.86	8	NAdam	9.22×10^{-5}	8.46×10^{-4}	0.012	0.192	0.045	0.073	93 141
c6a9839c	MobileNet V2	B	1.41	1.39	1.31	8	NAdam	4.85×10^{-5}	2.44×10^{-4}	0.011	0.266	0.044	0.075	91 855
2678f483	EfficientNet B4	B	1.02	1.88	1.01	16	NAdam	7.21×10^{-5}	2.30×10^{-4}	0.010	0.190	0.036	0.060	123 001

Table 11 – Summary of the hyperparameters and evaluation metrics of the three best performing deep learning models for each backbone in experiment phase 2.

Model name	Backbone network	Class loss weight	Bounding box regression loss weight	Flux regression loss weight	Mini-batch size	Optimizer	Learning rate	Weight decay	mAP50	Precision	Recall	F1-score	Flux RMSE (electrons)	Star centre RMSE (pixels)
f2c73ea4	ResNet 50	1.71	1.78	1.76	16	NAdam	8.95×10^{-5}	6.85×10^{-4}	0.631	0.938	0.634	0.757	38 469	0.675
42628b85	ResNet 152	1.49	1.50	1.43	16	AdamW	4.20×10^{-5}	7.94×10^{-4}	0.622	0.926	0.626	0.747	25 860	0.693
4d7a711f	ResNet 50	1.56	1.90	1.55	16	AdamW	5.47×10^{-5}	5.46×10^{-4}	0.622	0.941	0.625	0.751	22 723	0.670
0f128471	ResNet 152	1.78	1.52	1.38	16	NAdam	2.81×10^{-5}	1.81×10^{-4}	0.608	0.928	0.613	0.738	92 230	0.740
2be944d8	ResNet 50	1.16	1.35	1.81	8	NAdam	5.54×10^{-5}	1.02×10^{-4}	0.607	0.933	0.611	0.738	36 656	0.649
acc67ded	ResNet 152	1.63	1.80	1.18	8	AdamW	8.70×10^{-5}	9.25×10^{-4}	0.604	0.923	0.612	0.736	35 152	0.634

Note: Predictor architectures A, B and C correspond to the layouts *Input-Output, Input-512-Output* and *Input-512-128-Output* described in Table 8, respectively.

This page is intentionally left blank.

Table 12 – Average of the evaluation metrics for the three best performing models for each backbone

Backbone network	mAP50	Precision	Recall	F1-score	Flux RMSE (electrons)
ResNet 50	0.507	0.904	0.513	0.653	48 648
ResNet 152	0.472	0.887	0.483	0.624	51 460
DenseNet 161	0.089	0.485	0.127	0.194	101 923
MobileNet V2	0.040	0.391	0.084	0.138	86 077
EfficientNet B4	0.014	0.233	0.042	0.070	100 769

Regarding the hyperparameters tested, by analysing Table 10 it can be noted that some trends start to emerge that can reduce the hyperparameter space:

- Five of the six best models with ResNet architecture use the type B predictor architecture, i.e., an intermediate fully connected layer with 512 neurons and ReLU activation function is placed between the input and output layers.

- For the classification, bounding box regression and flux regression loss weights no clear pattern can be noted.

- Of the possible mini-batch sizes of 8, 16 and 32, the three best models of each backbone type use a mini-batch size of either 8 or 16, with 8 being the most common.

- Of the four types of optimizers tested, a clear preference towards the NAdam and AdamW solutions is visible, with NAdam being the most common by a large margin.

With the patterns found in the first experiment phase, the hyperparameter search space can be considerably reduced for the second experiment phase. Table 13 summarizes the new hyperparameter space that will be used in the next phase where three runs will be done per backbone for a total of 200 epochs each and with a batch size of 512.

The two ResNet architectures were kept since, although it seems that the extra complexity of the ResNet 152 might not be justified, the first phase only run for 30 epochs, being unclear whether with more epochs and further examples, the extra complexity could pay off. Regarding the learning rate, although all but one of the top performing models have seen their values range between 5×10^{-5} and 1×10^{-4}, it is expected that with only 30 epochs the faster learning model may be preferred even though it may be more unstable. For this reason, the learning rate interval was not changed, since with 200 epochs, slower learning models may prove more stable and better performing in the long run.

Table 13 – Summary of the hyperparameter search space for the second experiment phase

Hyperparameter	Type of distribution	Range of possible values
Backbone network	Random choice with equal probability	Resnet50, Resnet152
Predictor architecture	Random choice with equal probability	Input-512-Output
Class loss weight	Uniform	Minimum: 1 Maximum: 2
Bounding box regression loss weight	Uniform	Minimum: 1 Maximum: 2
Flux regression loss weight	Uniform	Minimum: 1 Maximum: 2
Mini-batch size	Random choice with equal probability	8, 16
Optimizer	Random choice with equal probability	AdamW NAdam
Learning rate	Uniform	Minimum: 1×10^{-6} Maximum: 5×10^{-5}
Weight decay	Uniform	Minimum: 1×10^{-5} Maximum: 1×10^{-4}

5.2. SECOND EXPERIMENT PHASE

The second experiment phase will perform a more extensive search on the more promising hyperparameters identified in the first phase, from which the final deep learning model to be used for result comparison will be selected. Additionally, the optimization of the DAOPHOT model will also be performed in this phase, where a number of hyperparameters will be tested to select the best combination for the final comparison.

5.2.1. DEEP LEARNING MODELS

As described in Chapter 5.1, the second experiment phase was performed on a subset of the hyperparameters previously tested and the results obtained are summarized in Table 11. The overall differences between the ResNet 50 and ResNet 152 architectures are still difficult to establish, even with an increased training time. Regarding the mAP, precision, recall and F1 score metrics, the top three models shown in Table 11 have very similar performances, with a slight edge for model *f2c73ea4* utilizing a ResNet 50 backbone on mAP. However, when looking at the flux RMSE, model *4d7a711f* has a significant advantage, with 22 723 electrons versus model *f2c73ea4* with 38 469. Similarly, this model has also a slight advantage regarding the RMSE of the star centres (among the top three), and displays a slightly superior precision (overall), with only a marginally lower recall. For these reasons, model *4d7a711f* will be the deep learning approach selected for the remainder of this study.

5.2.2. DAOPHOT

Although not sharing the same parameter and hyperparameter concepts as traditional machine learning models, the DAOPHOT algorithm also has some factors (referred to as hyperparameters for consistency hereafter) that can be adjusted. The threshold pixel value above which to considerer detection candidates, the critical separation distance from which two stars separated by less than that will be considered as belonging to the same group and the pixel size of the square area used to perform photometric measurements from which the flux is calculated are the three hyperparameters that will be tested. A fourth, the sigma of the PSF that will be used to construct the PSF model will be derived from the image using the process described in Chapter 4.7.1 and therefore doesn't need to be searched.

The threshold pixel value will be calculated as the standard deviation of the image pixel values multiplied by a threshold factor (TF), which can be interpreted as the number of standard deviations above which to detect sources. The critical separation distance will be calculated as the FHWM of the model PSF multiplied by a critical separation factor (CSF). The pixel size of the square area used to perform photometric analysis, also known as fit shape (FS) will simply be searched from a list of possible square side lengths [81]. Since the threshold factor was found to be the hyperparameter to which the results are most sensitive, of the total number of 18 experiments conducted, only the best two (as ranked by mAP50) for each threshold factor are shown in Table 14.

The best performing model (as ranked by mAP50), *0633db68*, was selected as the baseline for the remaining of this study. Both models with a threshold factor of 4 show a slightly superior mAP50 with a considerably higher precision. When the threshold factor is 3, the model starts identifying a considerably large number of noise pixels as stars (false positives), and when it takes the values of 5, a considerable number of the faintest stars remain undetected, highlighting how sensitive the DAOPHOT formulation is to this hyperparameter.

Table 14 – Summary of the hyperparameters and evaluation metrics of the DAOPHOT models in experiment phase 2

Model name	TF	CSF	FS	mAP50	Precision	Recall	F1-score	Flux RMSE (electrons)	Star centre RMSE (pixels)
0633db68	4	2	13	0.356	0.959	0.371	0.535	18 128	0.929
607b5d2d	4	2	11	0.356	0.964	0.368	0.533	27 386	0.902
8facdbba	3	2	11	0.315	0.574	0.548	0.561	22 581	1.020
5b4f63cf	3	1	11	0.312	0.571	0.545	0.558	22 842	0.987
ef8c25a9	5	2	13	0.271	0.995	0.272	0.427	21 010	0.884
153307ef	5	1	13	0.268	0.99	0.27	0.425	21 085	0.886

5.3. RESULTS ON THE SYNTHETIC DATASET

The final stage on the study of both the deep learning and classical approaches on the synthetic data is their evaluation on the test dataset. It consists of 256 previously unseen images, containing a total of 27092 stars of different size and brightness. The three main indicators to be assessed are the ability of the models to detect the maximum number of stars, and their photometric capabilities in accurately identifying thew centre of each star as well as the flux emitted by each.

5.3.1. DETECTIONS

Perhaps the most desirable characteristic for the models developed in this study, is their ability to improve upon the detection capabilities of existing methods. Difficulties in detecting faint stars or overlapping sources is one of the key shortcomings identified in classical techniques (as described in Chapter 2.1.5) and the area where most improvement can be achieved. Table 15 shows a high-level comparison between the overall performance of the two approaches for the synthetic test dataset.

The main difference between the two models lies on the recall or true positive rate (TPR). This metric shows that the Faster RCNN is capable of correctly identifying 63.7% of all existing sources, while the DAOPHOT could only achieve 37.3%. This can be translated in an extra 7139 stars being detected out of a total of 27092, allowing for many previously unknown stars to now be studied and analysed. Regarding the precision, or positive predictive value, of both models, Table 15 also shows that no significant difference is found. Despite the DAOPHOT model having a better performance when considering this metric, the slightly superior number of incorrect predictions made by the Faster R-CNN (both in absolute and relative terms), can be considered marginal in light of the large increase in recall. The mean average precision is also considerable higher for the Faster R-CNN network, being also more stable across different IoU thresholds.

Table 15 – Overall performance of the deep learning and classical techniques for a classification threshold of 0.5 and an IoU of 0.5 (for all metrics except mAP where each threshold is indicated) on the detection of stellar objects for the synthetic test dataset.

Model name	mAP50	mAP75	mAP90	True positives (TP)	False positives (FP)	False negatives (FN)	Precision	Recall
Faster R-CNN	0.633	0.570	0.28	17 245	1 152	9 847	93.7%	63.7%
DAOPHOT	0.360	0.220	0.01	10 106	389	16 986	96.3%	37.3%

When the number of detections is further broken down by stellar magnitude (see Figure 36), the picture becomes even clearer. The number of predicted stars is almost the same between DAOPHOT and the Faster R-CNN for magnitude up to 17, when considerable differences become more evident. The extra detections of the Faster R-CNN model are largely occurring for fainter stars (higher magnitude) where they become more difficult to distinguish from the sky background noise. This is precisely the area where algorithms such as DAOPHOT start showing some difficulties and where improvements can have a greater scientific impact.

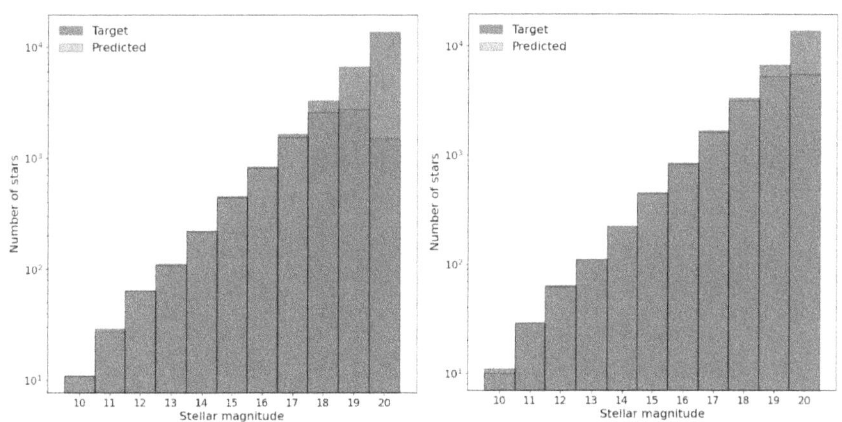

Figure 36 – Number of predicted and target stars (in logarithmic scale) as a function of stellar magnitude for the DAOPHOT model (on the left) and for the Faster R-CNN model (on the right).

This phenomenon can also be observed in Figure 37, where the two models are again very similar in their performance up to a magnitude of 17, above which the Faster R-CNN approach starts displaying superior detection capabilities.

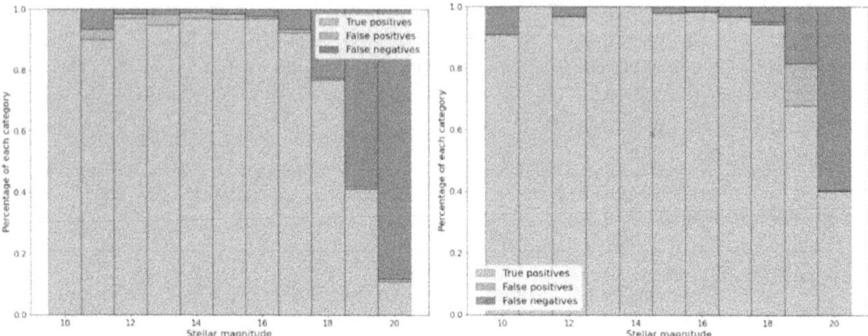

Figure 37 – Percentage of true positive, false positives and false negatives as a function of stellar magnitude for the DAOPHOT model (on the left) and for the Faster R-CNN model (on the right).

Further insights may be gained by looking at the performance of the two models considering how crowded or sparsely populated are the stellar fields. Clearly more pronounced for the Faster R-CNN, Figure 38 shows an increase in precision for this model as the number of stars per image grows, accompanied by a slight decrease in recall. One of the possible explanations may be tied with the fact that the network as a slight tendency to overestimate the number of stars present in the image, being too optimistic for sparsely populated images and more accurate for more crowded fields. For DAOPHOT the results are considerably more stable and do not show any significant changes as the number of stars per image varies. This image further highlights one of the critical differences between the DAOPHOT and Faster R-CNN approaches as being the considerable increase in recall.

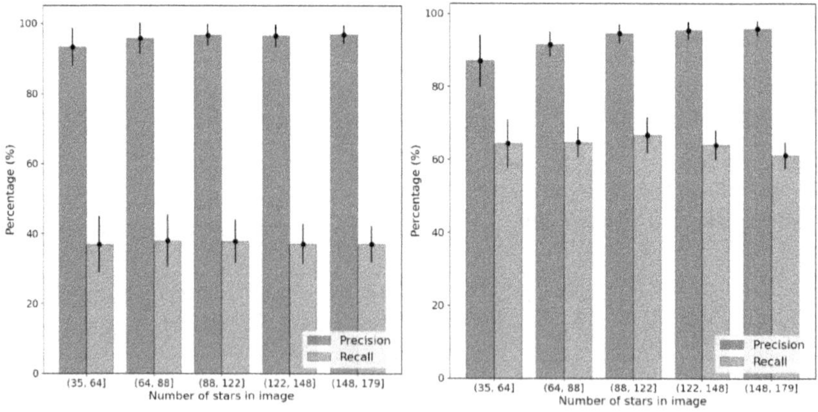

Figure 38 – Precision and recall as a function of the number of stars in each image (parenthesis identify closed intervals where square brackets represent open ones) for the DAOPHOT model (on the left) and for the Faster R-CNN model (on the right).

Finally, a more concrete understanding of the way each model works and the predictions they produce can be obtained by looking at individual image predictions. Figure 39 contains two example images from the synthetic test dataset where predictions have been made by both the DAOPHOT and Faster R-CNN models.

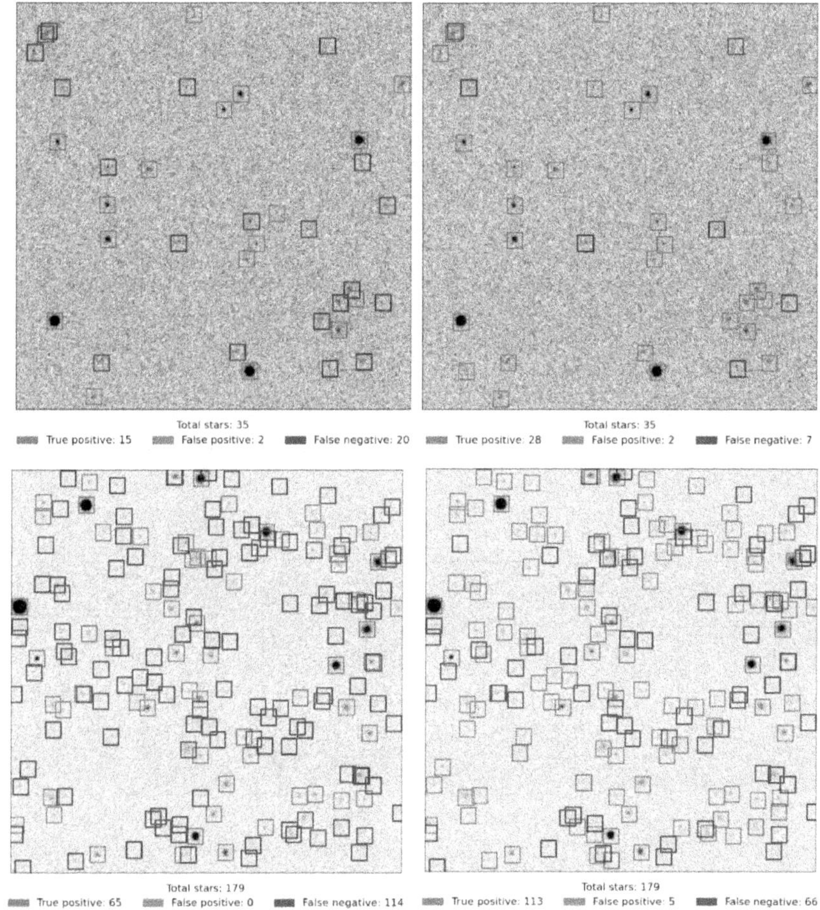

Figure 39 –Predictions of the DAOPHOT (left) and Faster R-CNN models (right) on two example images from the synthetic dataset. The image on the top row contains 35 stars representing a sparsely populated field and the image on the bottom row contains 179 stars highlighting more crowded conditions.

The first important remark can be seen in the considerable increase of true positives present in the images on the right-hand side of Figure 39. A transition from a majority of blue boxes (false negative – meaning the model considered the region to simply be background) of the images on the left to a majority of green boxes (meaning that the model correctly predicted the presence of a star) on the images on the right highlights the increase predictive power that is gained with the Faster R-CNN when compared to the traditional approach. Not only that but is it also clearly visible that (for this case and for most of the remaining images) all the stars detected by the DAOPHOT are also detected by the neural network, meaning that the extra correct predictions do not come at a cost of missing on previously detected stars. Further to that, this gain in new detections is mostly obtained for faint stars that are hardly visible to the naked eye.

Secondly, the Faster R-CNN approach displays good capability of detecting stars in close proximity with each other. This is particularly visible on the more crowded bottom right image of Figure 39, where many of the true positive detections have their boxes overlapping. Of course, there are also cases where two stars were so close that only one of them is detected, particularly if there's a large discrepancy in brightness between them. Nonetheless a clear improvement in detecting fainter stars in more crowded conditions can be seen between the DAOPHOT and Faster R-CNN images.

Figure 40 shows the number of detected stars spaced by less than 1 pixel, divided in 5 separation intervals. Both models were able to detect the same number (5) of stars separated by less than 0.2 pixels, however, for larger, but still sub-pixel separation distances, the Faster R-CNN network stars displaying a superior performance, being able to detect more closely spaced stars that its counterpart for all reimaging sub-pixel intervals. This gain in performance is also extremely valuable since many times, and particularly in crowded stellar fields, many stars are hidden in the vicinity of brighter neighbours.

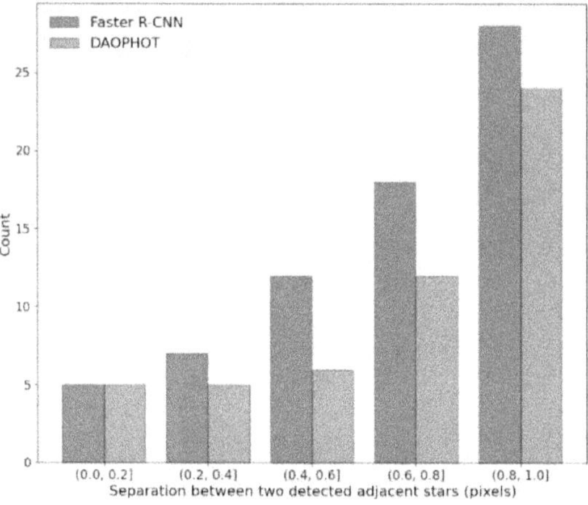

Figure 40 – Number of detected stars separated by less than 1 pixel per algorithm on the synthetic test dataset.

The final important remark is tied with the speed at which the 256 test images were analysed by each approach. The Faster R-CNN was able to run through all the images in approximately 65 seconds, reaching a speed of 0.25 seconds per image, while the DAOPHOT model need a total of 251 seconds to achieve the same task. This means a considerably slower speed of roughly 0.98 seconds per image, almost 4 times slower than the deep learning approach. This of course is not significant for such a small dataset, but when looking at the vast amount of data that is currently being generated by telescopes across the globe, may prove as a very beneficial gain.

5.3.2. STAR CENTRE

Although detecting more sources is a very important characteristic for the developed model, scientists and astronomers need further information about each star. One of the most relevant features is the exact location of its centre of intensity. The centre of intensity of a star is located at the centre of the theoretical PSF it produces. Since the number of photons originating from a given star and reaching the detector can be described by a Poisson distribution, the resulting shape of the PSF in the image may differ from the theoretical one.

DAOPHOT deals with this problem by approximating a given function (typically Gaussian-like) to various locations in the image and considering a star in locations where the error between the function and the real image data are beneath a given value. It can then use the known theoretical function to calculate all relevant properties, such as the position of its centre. For the Faster R-CNN model the location of the centre is determined by the centre of the bounding box surrounding each star. As the locations of the true theoretical star centres are known for each star in the synthetic dataset and the true bounding boxes are constructed from this starting point, the already present bounding box regression in the classical Faster R-CNN model is, by extension, also a regression to the coordinates of the star centre.

The average error for each model on its prediction of the centre coordinates of each detected star can be seen in Figure 41 for each magnitude band. The Faster R-CNN model achieves a smaller average error at all magnitudes despite the considerably higher standard deviation, particularly for bright stars (i.e., small magnitudes). It achieves a fairly precise measurement, averaging an error of around 0.25 pixels up to magnitude 17, from which the error starts growing until it reaches a maximum average of approximately 0.6 pixels for magnitude 20. The DAOPHOT model on the other hand shows a larger error on the stellar centre measurement averaging around 0.75 pixels up to magnitude 17, from where it grows to circa 0.9 pixels at magnitude 20. The standard deviation for this model is however generally smaller. For both models the precision on the measurement of the stellar centre decreases as stars get fainter. This is certainly not unexpected since when the pixel values of a star start approximating those of the background noise, smaller perturbations have a proportionally larger impact on the pixel distribution of these regions.

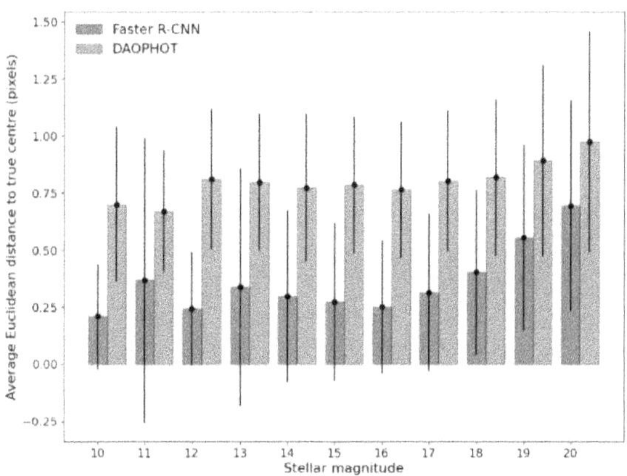

Figure 41 – Average Euclidean distance between predicted and true star centres per magnitude band on the synthetic test dataset.

Another way to visualize the differences between the precision of the prediction of star centres between the two models is by looking at the distribution of the errors obtained for all stars. Figure 42 highlights this, where the distribution of the Faster R-CNN prediction errors can be seen to concentrate considerably closer to 0 (peaking at around 0.25 pixels) and have a right or positive skew. The DAOPHOT model shows a much smaller number of precise predictions with its distribution having the highest density at around 0.9 to 1 pixel. Its skew is also much less pronounced than that of the Faster R-CNN.

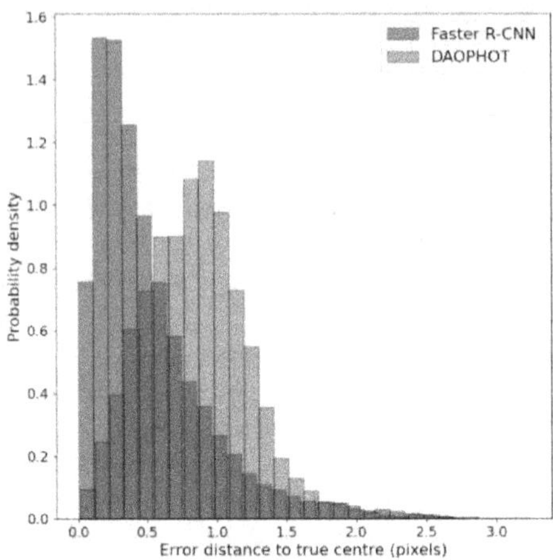

Figure 42 – Probability density distribution of the error Euclidean distances between the predicted and true star centres on the synthetic test dataset.

5.3.3. FLUX

Another key photometric attribute to be extracted from the images of stellar field is the flux of each star. For a point source, the flux can be extracted as the volume underneath the theoretical PSF that would represent each object, and calculating this quantity is exactly the method that DAOPHOT uses to estimate this value. The Faster R-CNN, as explained in Chapter 4.4.2, was modified to be able to also provide a flux estimate as an additional output. Only the predictive head was modified, meaning that the main body of the model, responsible for feature extraction and mapping was left unchanged.

The ability of each model to adequately predict the correct flux for each star can be first seen in Figure 43. This scatter plot shows how the proposals of each model differ from the target values for each star and allow for a more detailed analysis of which approach is performing better or worst. Starting with the DAOPHOT algorithm, and looking at the lower half of the image, we can see that for very dim stars, with very small fluxes, the model as considerable difficulties is performing an optimal fit to the data and overestimates many of these samples. As the target flux of the stars increases, DAOPHOT is then able to separate the star more clearly from the noise and the fit greatly increases leading also to a much more precise flux measurement. The image on the right, comprising now the Faster R-CNN, conveys an almost reciprocal story. For small values of target flux, the model is reasonably precise, being able to adequately fit the true data, however, as stars become brighter and their flux increases, the neural network stars to slightly underestimate its true value, giving rise to predictions that are slightly inferior to the expected values.

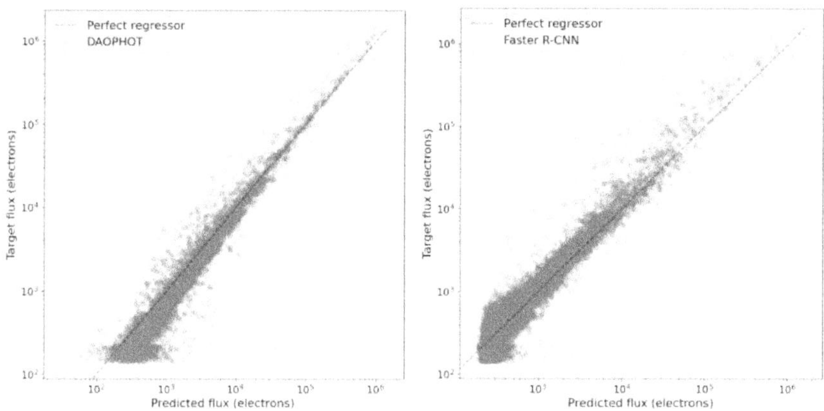

Figure 43 – Predicted and target fluxes for the DAOPHOT (left) and a Faster R-CNN model (right) on the synthetic test dataset.

These conclusions can be further validated in Figure 44, where the mean flux error and its standard deviation are shown for each magnitude interval. The logarithmic scale of the y-axis causes some distortion particularly for larger values but is still clearly visible that the Faster R-CNN produces larger errors for lower magnitude values, from magnitude 10, to 15. In this region, the DAOPHOT model seems to be more precise showing not only a smaller mean error, but also a decreased standard deviation. Moving into dimmer stars with larger magnitudes however, a different scenario emerges. As the stars get dimmer, DAOPHOT starts having more difficulties distinguishing the true sources from the sky background noise and its errors surpass those of the Faster R-CNN network for magnitudes above 15. This difference is even more pronounced for stars of magnitude 19 and 20, also where a significant number of stars reside.

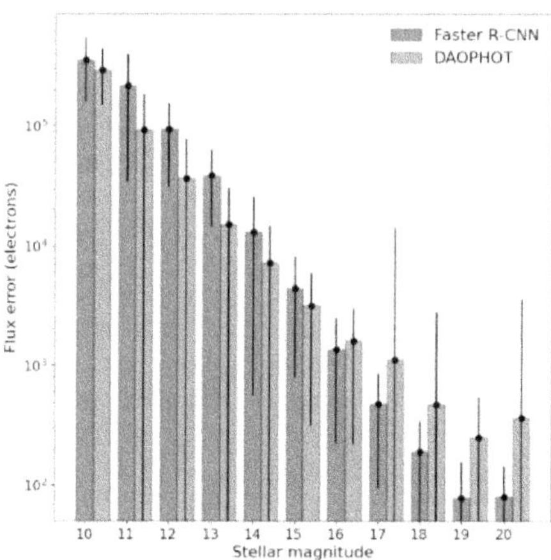

Figure 44 – Mean flux prediction error and its standard deviation per magnitude band on the synthetic test dataset for the DAOPHOT and Faster R-CNN models.

Finally, an overview of the distribution of the errors on the predictions of the two models can be seen in Figure 45. Both present an approximately normal distribution; however, the Faster R-CNN displays smaller errors overall with a significant portion of its shape being shifted to the left of the DAOPHOT model. Further to the that, the Faster R-CNN has also made more predictions overall, as can be seen by the difference in counts of the two distributions meaning that it was overall slightly more accurate while also making a larger number of predictions. The only exception can be seen on the bottom right corner of the image, where the number of predictions made by the neural network with significant errors (i.e., getting the flux wrong by more than 10 000 electrons and at times even 100 000) increases and overtakes the conventional approach.

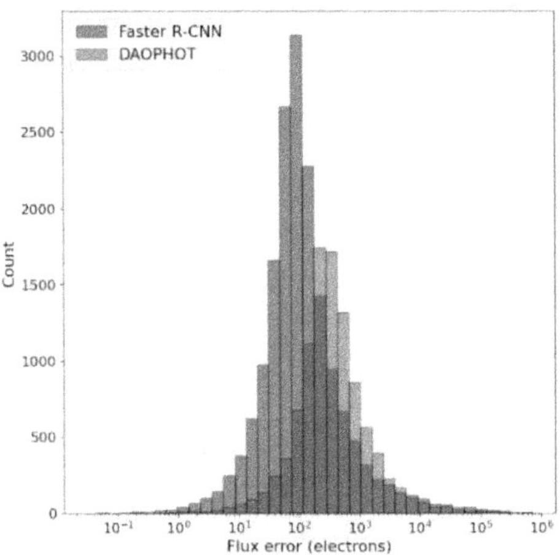

Figure 45 – Histogram of flux prediction errors for the DAOPHOT and Faster R-CNN models on the synthetic test dataset.

5.4. RESULTS ON THE COLLINDER 140 DATASET

Notwithstanding the good performance of the deep learning model on the synthetic dataset, it is also of critical importance that its performance on a real-world dataset is evaluated. Synthetic data can be extremely useful for training the model and running experiments in a controlled setting, where the ground truth is known, but the goal is naturally to make predictions on real image data.

5.4.1. ADAPTING FROM A SYNTHETIC TO A REAL DATASET

Although the synthetic data has proven very robust for training regarding detection and even the prediction of the star centre for real data, the same cannot be said regarding the flux prediction. Figure 46 shows a scatter plot relating the star fluxes estimated by the catalogue, with those predicted by the Faster R-CNN. It can be easily seen that the neural network underestimates the flux for most of the data points when comparing with the catalogue data.

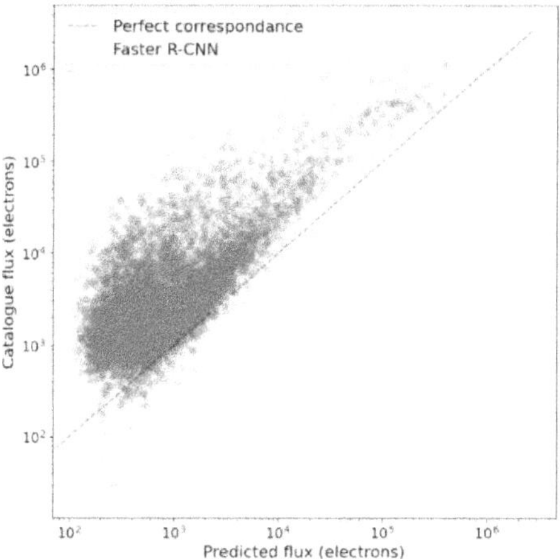

Figure 46 – Fluxes predicted by the catalogue and the Faster R-CNN (trained only on synthetic data) for the Collinder 140 test dataset.

Various approaches could be explored to address these discrepancies (see Chapter 6.4 for further discussion), however, the strategy selected in this study was to re-train only the neurons of the prediction head related to the flux regression on real Collinder 140 data. The remaining parameters of the model were frozen so that the network does not learn to mirror the predictions of the catalogue. This would be undesirable and in fact defeat the main purpose of training in synthetic data, which is to teach the network to look for stars that were previously unseen by the traditional methods. With only the 513 parameters (512 weights plus 1 bias), corresponding to the neuron responsible for the prediction of the flux output value being trainable, the network was trained in 768 train and validation images for a total of 20 epochs (as more epochs started having a detrimental impact on the validation results – showing initial signs of overfitting) and maintaining the remaining hyperparameters as per the second experiment phase with the exception of the learning rate which was lowered to 5×10^{-6}.

5.4.2. DETECTIONS

Differing from the case of the synthetic dataset, where the ground truth is known for each individual star (i.e., number of stars, star location and flux), in the case of the Collinder 140 dataset it is impossible to know the absolute truth. Better instrumentation and increased image resolution improves the likelihood that a given dim source becomes easier to distinguish from the sky background noise, however it is impossible to be certain that all stars are identified and there's none left to discover. For this reason, the analysis of the Collinder 140 dataset results will differ slightly from the one performed in the synthetic dataset. Although catalogue data produced by professional astronomers with access to well tested and established software is available, this catalogue results cannot be interpreted as the absolute truth about the Collinder 140 cluster, particularly since it is one of the key goals of this project to uncover new sources not yet detected by the traditional methods.

Table 16 contains a summary of the overall predictions made by the two models tested and compares them with the catalogue data. The Faster R-CNN model was able to reproduce the same predictions as the catalogue for 8507 stars, out of the 9629 predictions made by the latter. In other words, 88.3% of the predictions made by the catalogue were also made by the Faster R-CNN. The same number when considering the DAOPHOT drops instead to 74.5%. Conversely, only 1122 predictions made by the catalogue (or 11.7% of all predictions the catalogue made) were not proposed by the Faster R-CNN, while for the DAOPHOT, this number is instead 25.5%. Regarding the extra predictions made by each model, the neural network estimated a total of 34732 stars for all images (approximately 3.5 times more than the catalogue), while the DAOPHOT proposed a total of 12605 images, roughly 1.3 times as many as the catalogue data. Assuming a precision of 93.7% for the Faster R-CNN (as per the results on the synthetic dataset), this would mean that an extra 22915 stars would have been discovered by the neural network, in addition to those already identified by the catalogue. Although this results sound promising, it is nonetheless an ambitious assumption to admit that the precision would remain constant between datasets. Nonetheless, it is also important to mention that, admitting a stellar distribution as per the one described in Chapter 4.3.2, the total number of stars present in this image would be equal to 42112, which is much closer to the 34732 proposed by Faster R-CNN, than the 9629 proposed by the catalogue or the 12605 by DAOPHOT.

Table 16 – Overall performance of the deep learning and classical techniques for a classification threshold of 0.5 and an IoU of 0.5 on the detection of stellar objects for the Collinder 140 test dataset.

Model name	Coinciding predictions	Extra model predictions	Extra catalogue predictions
Faster R-CNN	8 507	26 225	1 122
DAOPHOT	7 175	5 430	2 454

Further insight may be gained by looking at the distribution of the number of stars by magnitude that are obtained for both the Faster R-CNN and the DAOPHOT models, as shown in Figure 47. The distribution of the DAOPHOT model more closely approximates that of the catalogue, where after a magnitude of 17 the number of stars stops growing exponentially with the increase in magnitude. This exponential increase is much clearer on the Faster R-CNN model, where a larger number of faint stars are detected.

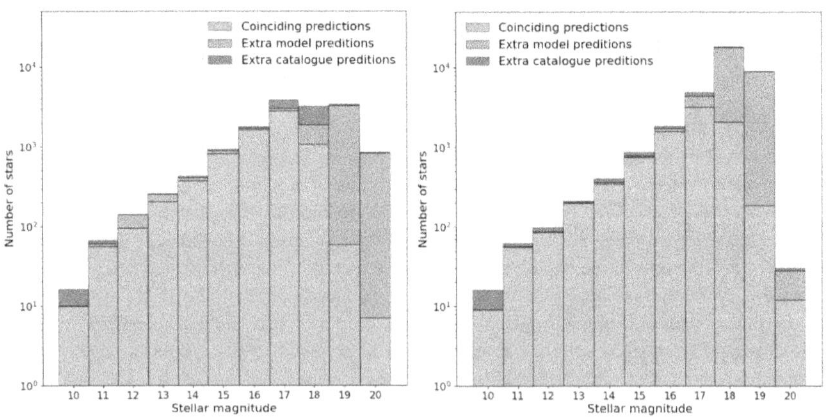

Figure 47 – Distribution of the number of stars (in logarithmic scale) per magnitude band for the DAOPHOT model (on the left) and for the Faster R-CNN model (on the right) on the Collinder 140 test dataset.

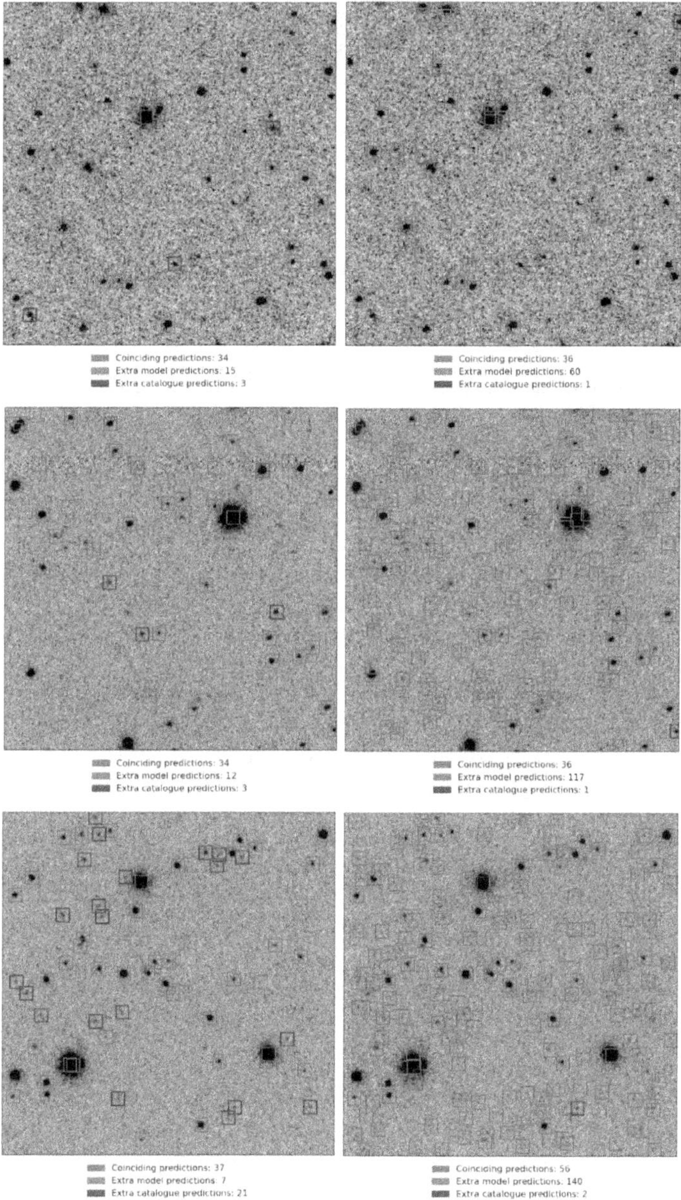

Figure 48 – Predictions of the DAOPHOT (left) and Faster R-CNN models (right) on three example images from the Collinder 140 test dataset, where each row corresponds to a given image predicted by each model. All images are clipped at three times the standard deviation of the sky background noise and shown in logarithmic scale.

Figure 48 contains the predictions of both the DAOPHOT and Faster R-CNN models on three example images selected from the Collinder 140 dataset to illustrate three different stellar field densities, from more a sparsely populated region on the top row to more crowded one at the bottom. It is possible to note for all images that the predictions of the DAOPHOT model tend to be closer to those of the catalogue, as they rely on similar principles, of course with some variations. Their predictions tend to match for brighter and better separated stars, however when the amplitude of the sources gets closer to the median sky background noise, the discrepancies become more visible between these two approaches. Several factors may change the end result for DAOPHOT, with one of the critical ones being the threshold factor discussed in Chapter 5.2.2. Smaller thresholds may lead to more detections, but of course with the side effect of incorrectly interpreting noise as a stellar point source.

As show in the images on the right, the perfomance of the Faster R-CNN differs more substantially from that of the catalogue. While almost all of the same stars are detected by both methods, the Faster R-CNN also detects a considerable number of fainter sources. In this region of pixel values, so close to the median sky background noise, it is very difficult by just using the human eye, to be certain whether or not these detections correspond to true stars, nontheless, the large overlap in predictions between the neural network and the catalogue provide additional confidence. Admitting these as true point sources, an extra 317 stars would be uncovered by the Faster R-CNN model, adding to the 132 already identified by the catalogue. This corresponds to an increase of 140% over the original known sources, which could completely alter the scientific knowledge we have of these regions of space.

5.4.3. STAR CENTRE

Unlike the synthetic test dataset, on the Collinder 140 dataset the true positions of the star centres are not known. It is impossible to know the absolute true value of this property, being the best option available to use the measurements of an established technique for comparison. Therefore, the results of the Collinder 140 catalogue will be used as the baseline for this metric, to which the two tested approaches will be compared.

Figure 49 shows the average distance between the predicted star centres and their location in the catalogue data per magnitude band. The first noticeable characteristic is the relatively large error for both the DAOPHOT and Faster R-CNN models on the objects with magnitudes 10 and 11. Stars with this magnitude are often so bright that the image may suffer from localized saturation, making the shape of the PSF differs more significantly from its theoretical form, and therefore the localization of its centre becomes more difficult to pinpoint precisely. For these magnitude bands, the two models perform roughly identically and even the position of the star centres suggested by the catalogue must be taken cautiously.

For dimmer, larger magnitude stars however, the results of the DAOPHOT model are considerably closer to the those proposed by the catalogue. The average distance from the catalogue centre for each band never exceeds a quarter of a pixel, being even considerably smaller in many cases. This proximity in performance is not unexpected as both models follow on similar approaches where a candidate curve is approximated to the distribution of the PSF in the image and positioned in such a way as to minimize the fit error.

The performance of the Faster R-CNN model differs more significantly from that of the catalogue. The average distance from the centre proposed by the neural network to that of the catalogue is typically between 0.75 and 1 pixel, for magnitudes larger than 11. This results in an inversion in performance from the synthetic dataset where the Faster R-CNN showed a better performance than the DAOPHOT. One of the main reasons for this may be tied with the fact that while on the synthetic dataset the shape of the PSF is based on the theoretical formulation of the two-dimension Gaussian curves, the shapes in the real Collinder 140 images show increased variation. Another likely factor is tied with the slight variation in the objects' position for each of the four image channels implemented in the synthetic training set. Only one sky observation was available for the Collinder 140 dataset, meaning that only one channel is used. Thus, the position of star centre used for comparison (as available in the catalogue data) is the actual location of the star in that image frame, and not its true position in the sky. Since any image is affected by atmospheric perturbations, that location will not correspond to its true location in the sky. This fact means that DAOPHOT able to achieve a very good performance on the Collinder 140 dataset, since what is being measured is the distance to the star centre in the image, and not to its true centre in the sky.

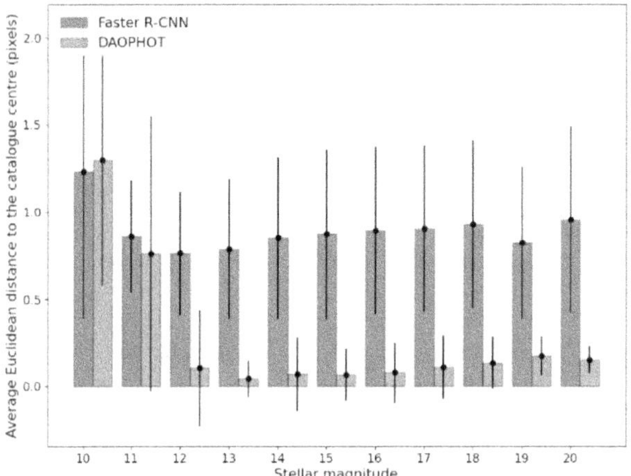

Figure 49 – Average Euclidean distance between predicted and the catalogued star centres per magnitude band on the Collinder 140 test dataset.

A full view of the distribution of the distance between the model prediction of the star centres and that of the catalogue can be seen in Figure 50. The vast majority of DAOPHOT predictions are within 0.1 pixels of the location where the catalogue places the star with virtually every prediction being within 0.5 pixels. The distribution for the Faster R-CNN is considerably more symmetrical and closer to a bell shape (albeit not exactly normal). Centred at around 0.75 pixels, it also shows a more pronounced tail to the right side, meaning that for some stars the difference can reach more than 2 pixels.

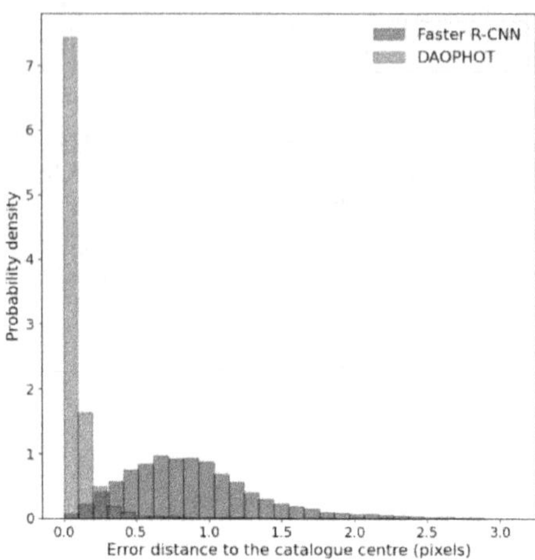

Figure 50 - Probability density distribution of the error Euclidean distances between the predicted and catalogued star centres on the Collinder 140 test dataset.

5.4.4. FLUX

As described in Chapter 5.4.3, it is impossible to know the exact value of the flux of each star in the Collinder 140 dataset. Once again, the best available option is to rely on the official catalogue data as the closest to the ground truth and use that as a baseline comparison to assess the performance of the two models.

Figure 51 shows a scatter plot comparing the flux predicted for each star by each of the models tested to that of the catalogue. Starting with the DAOPHOT results, we can see that the predictions closely match those of the catalogue for a vast majority of the samples, differing only in a few regions. For dim stars, most of the points plotted can be seen above the line of perfect correspondence, meaning that the model is estimating objects to have a lower flux than the catalogue expects. This is again visible for very bright objects (flux between 10^5 to 10^6) where DAOPHOT proposes a lower flux. The flux predictions of the Faster R-CNN differ more substantially to those of the catalogue. Although reasonably concentrated around the line of perfect correspondence, the spread of the points scatter around it is much larger than that of DAOPHOT.

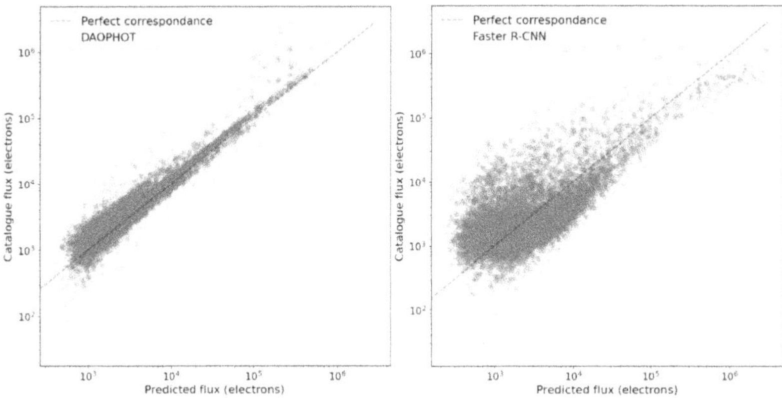

Figure 51 – Predicted flux for the DAOPHOT (left) and a Faster R-CNN model (right) compared to that of the catalogue on the Collinder 140 test dataset.

Another way to visualize these results is by looking at the mean error in flux prediction (assuming the catalogue values to be true) per magnitude band, as shown in Figure 52. In this image it is possible to note that the error for both models decreases with an increase in magnitude, i.e., dimmer stars with small flux values also have smaller errors. However, the errors obtained by the DAOPHOT model are always on average lower than those obtained by the Faster R-CNN for all magnitude bands. This, in conjunction with the scatter plots of Figure 51, highlight some limitations in the ability of the neural network to predict the stellar flux with the desired accuracy in the Collinder 140 test dataset.

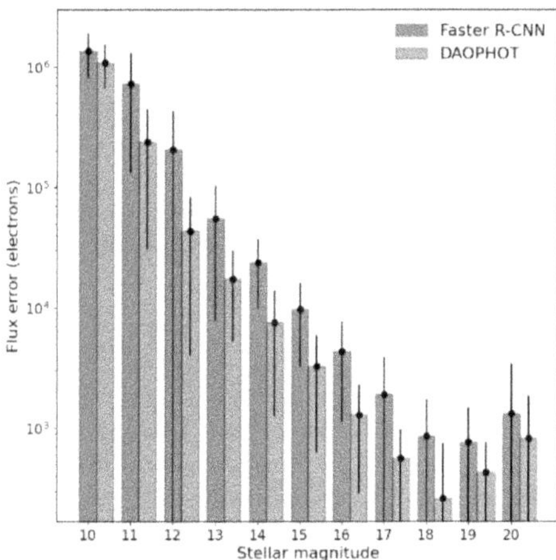

Figure 52 – Mean flux prediction error and its standard deviation (when compared to catalogue predictions) per magnitude band on the Collinder 140 test dataset for the DAOPHOT and Faster R-CNN models.

A final look at the full distribution of the prediction errors (assuming catalogue flux values as the ground truth) can be seen in Figure 53. Both DAOPHOT and the Faster R-CNN models show an approximately normal distribution, with the central value of the latter being just under an order of magnitude higher than that of the former. The reasons for this likely inferior precision in measuring the stellar flux by the Faster R-CNN may also be to a large extent due to the less idealised shaped of the stars in the Collinder 140 dataset as opposed to the synthetic data. Even though the flux prediction head was retrained on the real data (as detailed in Chapter 5.4.1), and improvements were achieved (see Figure 46 compared with the right image in Figure 51), the remainder of the network layers were frozen, thus not allowing the model to extract new features than those already computed in the synthetic data.

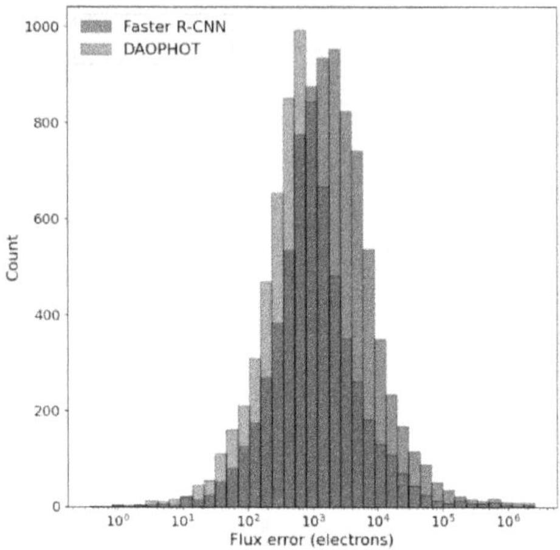

Figure 53 – Histogram of flux prediction errors (assuming catalogue values as true) for the DAOPHOT and Faster R-CNN models on the synthetic test dataset.

5.5. KEY REMARKS

The first experiment phase considerably reduced the possible space of promising hyperparameters. Backbones of the ResNet family and optimizers like AdamW and NAdam obtained results obtained considerably better results than their alternatives. This helped reduce the search space, allowing the second phase to be more focused on hyperparameter fine-tuning, yielding a further gain in performance.

Considering the results of both models, the Faster R-CNN achieved better performance than DAOPHOT on the synthetic data for all three tasks proposed (detection, localization, and flux prediction). On the Collinder 140 catalogue, the neural network coincided with the professional catalogue in a larger number of detections than DAOPHOT, also highlighting its good performance on real an unseen data regarding this task. Regarding localization, despite still with sub-pixel precision, DOAPHOT's prediction were closer to the catalogue data. Finally, the star fluxes predicted for DAOPHOT were again closer to those proposed by the catalogue, indicating that the deep learning model still has greater room to improve on these tasks.

6
DISCUSSION

6
DISCUSSION

6.1. DEVELOPMENT PHASE

The development of all the logical parts of the study can be divided into three main sections. The first concerns image generation and was responsible for producing the synthetic images that were used to train the models studied. The second part was focused on all the logic necessary to implement and train both the classical as well as the neural network model. Finally, the third and last part consisted of the post-processing necessary to evaluate the results, spanning from metric calculation to the production of graphics and analysis. This section will detail the key remarks found during development as well as identify any important findings that may guide future works.

6.1.1. IMAGE GENERATION

The process of image generation was originally developed using the *Numpy* [82] package for *Python*. However, since during training brand new images are being generated hundreds of times per epoch, this process quickly started getting too slow for an efficient development. As most modern deep learning models require a significant computational power, these are almost always trained by powerful graphical processing units (GPUs) that excel at matrix operations. However, since *Numpy* only operates on the central processing unit (CPU), data was constantly being created in the CPU and moved to the GPU for training, in a very slow and cumbersome process. To address this, the decision was taken to use *PyTorch* [77] tensors for the synthetic image generation directly on the GPU, thus eliminating unnecessary transfer of data. This change alone was responsible for a nearly 10-fold decrease on the training time of the deep learning network, going from around 5 minutes per epoch (512 images) to only just over thirty seconds.

The process for star generation consisted of the application of a two-dimensional Gaussian function (see Chapter 4.3.6) to a set on initial constrains regarding the star properties (namely, the centre coordinates, amplitude, and sigma). Further phenomena were considered to allow for a more realistic modelling of the expected pixel values, such as a Poissonian probability of each photo hitting the detector, however, all the stars generated followed this symmetrical Gaussian shape. Although the performance of the model was still extremely satisfactory, even when exposed to images that may not have exactly Gaussian PSFs (see the detection results on the Collinder 140 test dataset – Chapter 5.4.2), for other metrics such as the centre coordinates of a star and its flux, it is thought that the neural network could have benefited from being trained in a more diverse setting. Alternate formulation for the PSF, such as the Lorentzian [83], or the Moffat [84] could provide more variability and make the training more robust and adaptable to varying conditions. Future works could also benefit from using non-symmetrical formulations.

For the simulation of atmospheric conditions, such a sky background noise, and other defects that may affect the final image such as the behaviour of light passing through the detector of the capturing device, a fairly simplistic approach was adopted as described in Chapter 4.3.1. On the other end of the spectrum, approaches such as those described in PhoSim (Chapter 3.1) may provide incredibly precise estimations of all the physical phenomena light goes through from source to detection, of course at the cost of a large increase in complexity. Intermediate solutions, such as the formulations adopted by the SimCADO package [85] may offer the ideal compromise between complexity and a more precise representation, by offering a more accurate representation of the conditions at the time of observation, without an excessive increase in implementation and time complexity.

Notwithstanding the observations mentioned in the above paragraphs, the simplified modelling of both the shape of the PSF as well as the noise and defects to the images, was still able to achieve extremely promising results. As detailed in Chapter 5.4.2, a neural network with no further training than that performed on synthetic images generated, was able to identify 89% of the all the stars that were also identified by professional photometric analysis on the real-world Collinder 140 dataset. Although more accurate modelling of all the physical phenomena is thought to be beneficial, it is important to note that since relatively simple approach was already able to achieve very good results, and the extra complexity needed to perform more accurate generation may only yield diminishing returns.

6.1.2. MODEL DEVELOPMENT

The Faster R-CNN was the framework selected to be compared against the traditional techniques. Although it is thought to be the framework that had better chances of higher performance amongst the most common approaches, this selection is based on tests that are performed on general object detection and classification tasks (see Chapter 2.2.3). There is therefore a chance that other frameworks (e.g., regression/classification based or even novel frameworks) could have a superior performance for this specific task and may warrant further study.

Perhaps a more important task, with greater chances of resulting in a more significant performance gain, would be the construction of a bespoke framework. Architectures like Faster R-CNN and similar models are developed to be reasonably generic and have a good performance in a wide range of problems. However, they are not optimized to excel in any specific setting. The modifications done to the Faster R-CNN to allow it to perform one extra prediction (the star flux), although not trivial, are still far from transforming the network into a model specially conceived to address this specific task. It is thought that additional benefit could be gained from a more profoundly revised architecture where the features extracted by the backbone and FPN networks could be more directly used to inform the centre of intensity and flux regression as well as other photometric properties of the star, in a similar manner to the work described in Chapter 3.3.1, where more complex subnets are used for each prediction. The current Faster R-CNN implementation tested in this study relies on the information arriving at its 1000 input neurons for the three predictions made (i.e., object detection, localization, and flux prediction). A more complex system could experiment new architectures, where increased complexity of the prediction head (e.g., more neurons or layers), may allow for more complex features to be extracted. Another approach would be for each prediction head to be derived directly from the Multi Scale RoIAlign step (see Figure 34). Instead of each task sharing the same 1000 input neurons (meaning that the networks' parameters cannot be optimized for each task), each prediction could derive its information individually allowing for a more optimal parameter space to be reached.

Another key point on the development of the neural network model was the selection of the loss function. The loss function originally proposed for the Faster R-CNN consists of a cross-entropy loss for object classification and the smooth L1 loss for regression of the bounding box coordinates. These two functions were kept for each of the corresponding tasks, and a new smooth L1 function was introduced for the loss concerning the prediction of the star fluxes. However, more extensive research into this topic could prove extremely valuable. The loss function used by the standard Faster R-CNN model [42] was developed predominantly for object classification and regression to the coordinates of the bounding box. For this problem, the coordinates of the centre of the star (derived from the bounding box), as well as the flux of the star, assume a more critical role. Further experiments with different loss functions, be it the relative weight of each individual loss, the aggregation measure used (i.e., sum, average, etc.) and even the functions employed to each parcel could all bring further benefits to the model performance.

Despite the opportunities for future development mentioned in the previous paragraphs, extremely solid results were achieved by adapting the existing generic architecture of the Faster R-CNN network to this problem. With only minimal training time (generally under 2 hours for 200 epochs), and resources (a single GPU with 16 gigabytes of memory was available), it was not only possible to generate all the 100 thousand images needed for training as well and conduct train and validation at each epoch. By the end of the training process, the network was capable of not only achieving considerably better performance on previously unseen synthetic test data than the traditional methods, but also to propose many previously undetected stars on the real-world data.

6.1.3. POST-PROCESSING

Post-processing techniques, namely non-maximum suppression, are used in the Faster R-CNN model to sieve through multiple overlapping boxes that have a good correspondence with the underlying object. They are extremely useful in ensuring that only the best boxes with the highest chances of successful correspondence are kept while the remaining ones are excluded. Although this is generally satisfactory for most object detection problems, in this task, overlapping stars that lay very close to each other may get excluded in the process. Techniques such as Soft NMS [86] introduce a decay to the score of the remaining overlapping boxes once the most promising candidate is selected, instead eliminating them, as is the case for the greedy approach of classical NMS. This may prove more suitable to problems where the separation of nearby objects assumes a more critical role, as is the case in this study.

Future works could also perform post-processing in critical parts of the image, e.g., where there are "dead" or "hot" pixels or columns a larger confidence score could be adopted to classify a given prediction as valid. It could then be lowered for the remainder of the frame, where the image quality is known to be higher. This idea could be applied to both the neural network and classical models, allowing for the models to be more adaptable and respond more adequately to different environments. Although this approach may lower the total number of detections, it would ultimately further contribute to increase the confidence in the detected sources.

6.2. DETECTIONS

6.2.1. SYNTHETIC DATASET

Detection results on the synthetic dataset show a superior performance of the Faster R-CNN model compared to that of DAOPHOT. The neural network is not only capable of detecting more stars overall but does it mainly for dim stars where their amplitude is closer to the pixel values of sky background noise. This is a region where traditional models are known to face increased difficulties, but also where majority of stars are present [87]. The number of stars increases exponentially with the magnitude increase, of course not indefinitely, but the point at which this relationship breaks is thought to be beyond a magnitude of 20 (the maximum value analysed in this study). This stellar distribution was also noted in the Collinder 140 dataset and at the galactic centre of the Milky Way [73].

Albeit in a synthetic and controlled setting, the ability of the Faster R-CNN to correctly detect 64% of all existing stars, versus only 37% for the DAOPHOT, shows the potential that this new approach has in expanding the boundaries of what is possible to achieve with current technology, and this is without accounting for some of the possible improvements proposed in Chapter 6.1 which could raise the difference even further.

Not only the total number of existing stars correctly identified is relevant. Only 6.3% of all the stars predicted by the Faster R-CNN were background noise, a number very close to the 3.7% achieved by DAOPHOT. A reduced number of false positives is a very important characteristic for models of this kind, as it allows scientists to have increased confidence that the predictions can be trusted when the ground truth is not known, as is the case for most real-world datasets. Of the 1152 of stars incorrectly predicted by the neural network, 1127 (97.8%) had a magnitude of either 19 or 20, where the amplitude is typically beneath 100 electrons and often even below 30 electrons (well within the range of sky background noise). The results for DAOPHOT are also very robust regarding the number of false positive detections. Only 389 were predicted in total with 173 (44.5%) being within magnitudes of 19 and 20 (48 and 125 respectively). They, however, spread further into lower magnitude values when compared to the Faster R-CNN with 55.6% of false positives being between the magnitudes of 11 and 18, while for the Faster R-CNN this number was only 0.2%.

Regarding the number of false negatives, or real stars that were not predicted by the models, the results of the Faster R-CNN show that 9847 stars (36.3% of all the stars present) were missed. Of these, 8188 (or 83.2% of all undetected stars) had a magnitude of 20, and 1418 (14.4%) had a magnitude of 19, resulting that 97.6% of all stars undetected by the Faster R-CNN lie in just these two magnitude bands. These results are considerably more promising than those obtained by DAOPHOT where 16986 stars (or 62.7% of all stars present) remained undetected. Their distribution was also more spread over lower magnitudes with 20 stars at magnitude 16, 113 stars at magnitude 17 and 748 at magnitude 18 remaining undetected (0.1%, 0.67% and 4.4% of DAOPHOT's undetected stars, respectively). At the magnitude of 19, DAOPHOT missed on 3901 stars (approximately 2.2 times more than the neural network) and at magnitude 20 the classical method missed on 12189 stars, almost 1.5 times worse than the Faster R-CNN.

6.2.2. COLLINDER 140 DATASET

Considering the model performance on the Collinder 140 test dataset, the results of the Faster R-CNN are still more promising than its competing approach. Here, the absolute ground truth is not known, but professional catalogue data can be used to estimate with increased confidence the performance of each approach, particularly for lower magnitude values. Regarding coinciding predictions, i.e., predictions made both the catalogue and each model, the Faster R-CNN accrued a total of 8507, against 7175 by DAOPHOT. Comparing with the 9626 stars proposed by the catalogue, this means that the Faster R-CNN was able to detect 88.4% of all the stars also detected by the catalogue, while this figure for DAOPHOT remained at 74.5%. These results are extremely encouraging and show how the neural network, despite being trained in images that never existed and are simply and idealized representation of our understanding of the universe, is still able to apply that learned knowledge to real-world scenarios, compete and even surpass the most established techniques in the field.

Focusing on the catalogue predictions that were not made by the models, the Faster R-CNN network also achieves the best performance. It only failed to identify 1122 objects that the catalogue classifies as stars, while this number for DAOPHOT raises to 2454, equating to 11.6% and 25.5% of all the objects proposed by the catalogue. The distribution however is not as concentrated on high magnitude values as it was with the synthetic data. 805 (71.7%) of the proposed catalogue stars that the Faster R-CNN failed to agree on are situated at magnitudes 17 and above, leaving a still sizeable 28.3% for objects at brighter magnitudes. For DAOPHOT the picture is slightly different with 2311 (94.2%) of the missed detections being for object with magnitudes higher than 17, and only 5.8% beneath it. DAOPHOT therefore shows a slightly better performance for brighter and more visible stars, albeit at the cost of less detections overall.

The most revealing feat can be seen by looking at the extra predictions made by the models which the catalogue could not identify. Here, the Faster R-CNN was able to identify an extra 26225 stars with 1163 (4.4%) having magnitude 17, 16063 (61.3%) magnitude 18, 8804 (33.6%) magnitude 19 and 16 new stars at magnitude 20, resulting in 99.3% of the proposed new stars all being situated at magnitudes of 17 and above. These results further reinforce the key advantage of this method, allowing researchers to see further into regions of space that were previously too dim to be adequately mapped and surveyed. Comparing with the results obtained by DAOPHOT, the classical method was able to propose 5430 stars undetected by the catalogue, of which 165 (3.0%) at magnitude 17, 831 (15.3%) at magnitude 18, 3177 (58.5%) at magnitude 19, and 819 (15.1%) at magnitude 20, resulting in of 91.9% of its predictions being at or above the magnitude of 17 and 8.1% for brighter stars. It is however important to note that the exact magnitude that is assigned for each star relies on an accurate prediction of the star flux and, as will be seen in Chapter 6.4, the fluxes predicted by the Faster R-CNN on the Collinder 140 test data are still suboptimal when compared with the other methods.

Another important metric to account for is the total number of stars in a 256x256 pixel grid predicted by each approach. Staring with the catalogue data, each image is expected to have 37.6 stars on average with a standard deviation of ± 10.4 stars, resulting in an average density of 5.74×10^{-4} stars per square pixel. For DAOPHOT, these number would increase slightly to an average of 49.2 ± 17.9 stars per image or an average star density of 7.51×10^{-4} stars per square pixel. Finally, with the results of the Faster R-CNN approach, the increase would be significant, with each image now averaging 135.7 ± 46.2 stars, resulting in an average star density of 2.07×10^{-3} stars per square pixel. This would result in a coefficient of variation of 0.28, 0.36 and 0.34 for the catalogue, DAOPHOT and Faster R-CNN approaches respectively.

Notwithstanding the promising results obtained by the neural network on the Collinder 140 star cluster, it is important to also highlight some of the shortcomings identified in this method. On the top quarter of the middle image in Figure 48, it is possible to note a horizontal band, likely caused by stitching of adjacent images of different detector chips. This anomaly caused an abnormally large prediction of stellar objects in the region, which is unlikely to be caused by true sources. Additionally, on the bottom right corner of the bottom row image in Figure 48, a particularly large and saturated source cause the Faster R-CNN model to propose four additional predictions (in a cross shape) which are very unlikely to be due to true stars. These two examples highlight some of the challenges that the neural network still faces and that could be further expanded in future works, likely greatly improving its performance. One of the main ideas that could be attempted to improve on this would be to train the network in a more diverse set of images. Adding a range of image defects, such as the previously mentioned "hot" or "dead" pixels, stitching regions, etc., to the synthetic training set, could allow the model to recognize these regions more easily and better adjust its predictions to them. Further post-processing steps mentioned in 6.1.3, could also be used to further refine the proposals in these areas by either increasing the necessary confidence for an accepted prediction.

6.3. STAR CENTRE

6.3.1. SYNTHETIC DATASET

With the known true values coordinates of each star in the images, due to the synthetic generation process, the evaluation of both models in this context is a very useful indicator of the potential localization ability of each. The Faster R-CNN was able to achieve a superior performance on this data, with an average error of 0.52 pixels and a standard deviation of 0.43 pixels. The same metrics for DAOPHOT show an average of 0.85±0.38 pixels. These results show a better capability of the neural network to adequately pinpoint the centre pixel coordinates of each star, even though its errors tend to fluctuate more. The slightly higher stability of DAOPHOT errors, on the other hand, is still not sufficient to result on a more accurate measurement of the stellar centre of intensity.

Breaking down these results per star magnitude, a more detailed picture of their behaviour can be established. The Faster R-CNN shows a relatively low error up to magnitude 17, with an average of 0.29±0.34 pixels, while on larger magnitudes this figure grows to 0.57±0.44 pixels. This difference in performance is unsurprising, as dimmer objects are considerably more difficult to detect and their peaks more likely to be confused with sky background noise. This is an important consideration since in the theoretical formulation of the undisturbed PSF, its highest value would also correspond to the centre of intensity of the object. Given that the final value of each pixel in the PSF is taken from a Poisson distribution, where the rate parameter is that same original pixel value, the actual position of the largest pixel value on the final PSF, as captured in the image, may differ substantially from the position of its centre of intensity.

Although impossible to be certain, it is likely that since the neural network can learn from all the data, after sufficient training they will uncover a method to minimize the average error caused by the gap between the Poissonian PSF pixel values and the star centre. For DAOPHOT however, this phenomenon is thought to be more problematic. Since the algorithm relies on placing an idealized PSF at each position and measuring the error of fit to the actual image data, it is likely that this algorithm is less adaptable and more prone to these variations. Its results seem to corroborate this idea, with the errors from the true star position up to a magnitude of 17 having an average of 0.79±0.30 pixels, and larger magnitudes rising to 0.88±0.41 pixels. The increase here is not as pronounce as with the neural network, however it is overall further away from the desired values.

Subpixel estimation [88] focuses on approximating the value of a geometric quantity, in this case the centre coordinates of a star, to a higher level of precision than the quantized pixel space at which the data was sampled. Although the resolution at which the image was collected may be a limiting factor, some techniques may be employed to help minimize this restriction. One of the easiest approaches in the context of the Faster R-CNN, would be to increase the number of proposed boxes, by reducing the stride that separates them. This would generate a finer mesh of anchor boxes which could then be evaluated for the presence of an object. A more precise matching could be obtained with the underlying objects which could lead to an increased precision on the localization of the star centre.

Another possible technique would be to up-sample the original image fed to the network. Here, the idea would be to provide the network with a denser pixel mesh as to reduce the negative effects of the original sampling space on the precision of the object coordinates. Several techniques could be employed to estimate the subpixel values such as interpolation, integration, Taylor series approximation or phase correlation [89]. The network would then be trained on higher resolution images, artificially reducing the sampling space, and its predictions could be translated back into the original image size, resulting in more precise regression values at a subpixel scale. This, of course, would have the drawback of increased computational costs as now the network would have to be trained on images of larger size.

Additional investigation would have to be performed to evaluate the potential of each technique for lowering the error on the centre coordinates of the stellar objects. Despite the errors being typically under 1 pixel for most of the detected objects on both the synthetic and Collinder 140 datasets, an extremely accurate prediction is of upmost importance for scientists and astronomers, since when testing physical theories that explain our universe, the interactions between different bodies are highly dependent on their mass and location.

6.3.2. COLLINDER 140 DATASET

Looking at the performance of both models on the Collinder 140 dataset, a distinct picture from the synthetic test data emerges. First, it is important to highlight that for in this dataset, the absolute ground truth is not known, only the catalogue predictions. Therefore, what is being measured is not the distance of each model to the true coordinates of the centre of each star, but rather the error each model is able to achieve between their proposed coordinates and those proposed by the catalogue. Both models show a very large error for stars of magnitude 10, with an average value around 1.3 pixels and standard deviations of more than 1.5 pixels. This is because stars of this brightness exceed the full well capacity of the individual pixels, causing them to saturate. The result is a series of adjacent pixels forming a plateau, i.e., all having the same maximum value, which makes it very difficult for DAOPHOT to adjust a PSF of Gaussian shape, and for the neural network to estimate the theoretical centre of this shape. This problem is still present at a smaller scale for stars of magnitude 11 but is no longer visible for stars of larger magnitude.

For stellar objects of magnitude 12 and above, DAOPHOT achieves extremely robust results. The average error between the centre coordinates precited by this method compared with the catalogue, excluding magnitudes below 12, is only 0.10 pixels with a 0.18 pixels standard deviation. Further breaking down these results we can observe that for magnitudes between 12 and 17, where stars are still easily distinguishable from the sky background noise, the results are 0.09±0.18 pixels, rising to 0.13±0.15 for magnitude above 17. Despite the promising results and apparent great performance in both brighter and dimmer stars, it is important to highlight that these results only include detection share by both DAOPHOT and the catalogue. Therefore, the number of stars between a magnitude of 12 and 17 is 5975, while above it only 1130 are present, meaning that the result may suffer from a slight skew towards easier to detect stars whose shape is also closer to the theoretical formulation.

Another important fact to highlight is that the methodology that DAOPHOT and the catalogue use to estimate the star centre is based on similar theoretical formulations and thus may be prone to suffer from similar biases. Both algorithms rely on adjusting an idealised PSF to a given region of the image where a star is thought to be. In the fitting process, these methods may produce very similar outcomes when placed with comparable underlying pixel values even if these are of a shape distorted enough to produce significant errors from the true centre of the star.

The errors of Faster R-CNN on the localization of the star centre for the Collinder 140 dataset are less promising than those of the traditional algorithm. The network achieved an average error of 0.90±0.48 pixels from the centre positions proposed by the catalogue for stars above a magnitude of 12. Its errors are also relatively stable across the different magnitude intervals with this figure for magnitudes 12 to 17 being 0.89±0.47 pixels and above magnitudes of 17 being 0.92±0.48 pixels. Although these errors from the catalogue predictions are considerably higher than those of DAOPHOT they are more comparable, albeit still worse, to those obtained in the synthetic dataset where the mean and standard deviation for all stars were 0.52±0.43 pixels, respectively.

The main reason thought to contribute to a worse performance on the Collinder 140 dataset against the synthetic data is tied with the fact the neural network was trained on four channels. These channels aimed to represent atmospheric effects that cause the star position to vary slightly in the sky between sequential observations. However, since on the Collinder 140 data, only a single channel is available, the error measured in this case is not the error to the true star position in the sky (as it was for the synthetic data), but rather the error to the star centre position in the image. This also explains the significant gain in performance of DAOPHOT, and the similarity between its results and those of the professional catalogue data, as these methods are designed specifically for this task.

Another factor that may also contribute for this performance decrease tied with the variability of shape of the stars in which the model was trained. Figure 54 shows an example of two stars of similar brightness, one generated by the synthetic process and the other extracted from the real Collinder 140 dataset. They both correspond to correct detections made by the Faster R-CNN model. Although it is possible to see some similarities via a simple naked-eye observation, some more subtle differences are still noticeable between the two pictures. A more robust image generation process that could more closely approximate all the different shapes and forms that the real stars could take (possibly including other formulations in addition to the Gaussian), would have the potential to both improve and provide more confidence in the localization capabilities of the neural network. The prediction the network makes relies on the information contained in a critical subset of pixels (as illustrated in Figure 54), so it is of critical importance that these regions are characterized as realistically as possible.

Several techniques can be utilized to improve on the image generation process and produce outputs closer to those found in real observations. One of the simplest consists in using image similarity measures to quantify the degree of similarity between the generated and real images. Metrics as simple as root mean squared error, to more elaborate ones such as the structural similarity index (SSI), which makes use of the luminance, contrast and structure to evaluate the differences between two images [90], the feature base similarity index (FSIM) which further incorporates measurement of the local structure of low-level features in the image [91], or Fréchet inception distance (FID) which compares the distribution of the generated images with that of real ones [92]. All this metrics could be used to guide the image creation process and adjust the variables responsible for controlling the generation of synthetic images.

Depending on the result of these initial strategies, more complex approaches could be attempted, such as the use of generative adversarial networks (GANs) [93]. A discriminant network could be trained to distinguish between the real images and those that were synthetically produced. This setting could be trained unit the generation process was sufficiently robust to ensure that the discriminant could no longer set apart the images originating from both sources. That would provide increased confidence that the synthetic data accurately represents the expected real-world images and that the errors on the distance to the true star centre on the synthetic data would be a good benchmark for the precision that would be possible to obtain in real-world datasets.

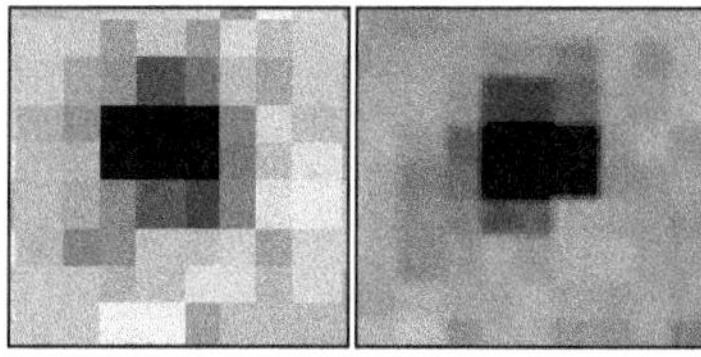

Figure 54 – Example of a star generated by the synthetic generation engine (on the left) and extracted from the Collinder 140 images (on the right)

6.4. FLUX

6.4.1. SYNTHETIC DATASET

The prediction of the flux value for each star shows a similar performance between the two models, with a slight edge on the overall results for the Faster R-CNN. The neural network was able to achieve an average absolute error on the predicted flux of all stars of 1628 electrons, with a standard deviation of 16902 electrons. The proportionally high standard deviation can be easily explained by the fact that bright stars have exponentially larger flux but are exponentially less common. The same results for the DAOPHOT show an average absolute error of 1791±14514 electrons. Not a significant difference but with a slight edge for the neural network on overall smaller error, and a slight edge for DAOPHOT on more stable predictions.

The flux errors, as expected, vary significantly with the stellar magnitude. For the brightest stars of magnitude 10, the best performing method is actually DAOPHOT with an average error of 293±143 thousand electrons, versus 355±197 thousand electrons for the neural network. For stars up to magnitude of 15, the scenario is essentially the same, with conventional algorithm having a slight edge over the Faster R-CNN, however, when the stars start getting dimmer and their magnitude raises above 15 a different picture stars to emerge. DAOPHOT stars having difficulty in accurately predicting the flux of a stars with its errors no longer decreasing exponentially with a magnitude increase. This is likely caused by the fact that as the amplitude of the stars approximating the values of the sky background noise, these noisy pixels star interfering more with the star shape and the fit of the PSF curve becomes flattened, resulting in a wider curve with larger flux. Considering only stars with magnitudes of 16 and above, the difference in performance is clear with the Faster R-CNN having an average absolute error of 205±413 electrons, while for DAOPHOT the same result reaches 596±5689 electrons. This is almost three times the mean of the former and more than 13 times its standard deviation, highlighting the difficulties faced by the algorithm for objects in this magnitude range.

6.4.2. COLLINDER 140 DATASET

The differences on the flux predicted by the Faster R-CNN on the synthetic dataset and later or the Collinder 140 data are the most distinct of the three metrics evaluated in this study. So much so, that the flux prediction head of the neural network was re-trained on a set of the Collinder 140 dataset (using the catalogue predictions as the true values) to improve on its performance (see Figure 46 and Figure 51). This introduces an additional step for utilizing the neural network directly as a pre-trained model on a custom dataset. However, as only 513 parameters (512 weights plus one bias) are being retrained, the extra step does not introduce significant obstacles and is still worth the extra performance gained. Nonetheless, it is still important to discuss why the performance decreased so critically for this task and if something could be done in the training stage on the synthetic data to prevent the need for the extra training.

One of the most likely causes for the difference in performance between the two sets of data is insufficient accuracy in the representation of the shape and form of the star on the synthetic image. Much of the discussion in Chapter 6.3.2 regarding the generation of image shapes closer to those available in real images could have a beneficial effect on the performance of the model transitioning from synthetic to real data. Additionally, the architecture of the Faster R-CNN itself is thought to be another of the key factors. It was originally designed for a more generic task where the key goals are the detection and localization of objects of different classes. However, the problem presented in this study differs slightly from the original goal, by the fact that sub-pixel precision as well as regression concerning the properties of the detected objects assume a more central role. The approach taken was to modify the predictive head to accommodate for these extra requirements, but it thought that a more profound alterations of the mechanisms of the network could have a more profound effect. Making use of the features extracted by the backbone and FPN, or even extracting custom features related to the flux and brightness of these objects that could be later used by the final layers could make better use of the available information and result in a model that can better approximate all the seen data.

7
CONCLUSION

7.1. KEY REMARKS

This study aimed to identify the advantages provided by state-of-the-art deep learning approaches over traditional photometric algorithms on the key points: the detection of stellar sources, the correct localization of their centre coordinates, and accurate prediction of their flux. Supported by the evidence highlighted in the analysis of the results obtained, it can be concluded that the neural network has a superior performance in object detection in both synthetic and real-world images. Not only is the performance superior but the extra objects encountered over the traditional methods, but these newly found stars are also of the magnitudes that they were expected to have, showing further consistency with the proposed research objectives.

Regarding the remaining research goals, the performance of the deep learning model was found to be superior in predicting the centre of the stellar objects and their flux for the synthetic dataset. However, when exposed to the real-word data, there are still some doubts concerning its achievements and whether it can provide more trustworthy results that the conventional technique in its current state. Nonetheless, the superior performance on the synthetic images highlights the potential that this approach has and shows that an improved generation process capable of producing more realistic images can lead to increasingly better results on real data.

The methodology developed allowed for an efficient and effective development process, with fast training times and adaptability to obstacles found during the process. It also permitted that, at each experiment conducted, the results obtained by deep learning approach could be compared with the best algorithms currently used in the industry, establishing continuous a baseline for the study and ensuring that all the important features could be considered. This strategy also allowed to reduce the search space of possible development strategies quickly and effectively, by first reducing it to a small set of most promising hyperparameters and then finding their ideal combination.

Finally, this book helped develop the stepping-stone from which future works can improve upon and focus on the more specialized parts defined herein, instead of having to tackle to full problem at once. The key contribution of this work is to define the strong points of this approach and shed light on the more problematic areas that still require further studies. It also helped establish the importance of having a synthetic generation engine that can not only guide the development process, but also help in understanding the idealized potential of this method. Further to that, it proved that with a generation engine of relatively simple formulation, accurate representations of stellar fields can already be achieved, helping focus the objectives of future works where the most important improvements can be obtained.

7.2. FUTURE DEVELOPMENTS

While several recommendations for further developments on this task were mentioned throughout the study (particularly in Chapter 6), this section aims to summarize the paths that are thought to have the most potential to help address the main shortcomings found. Although the results on the detection of new stars, and particularly of faint sources, are extremely promising and point to the enormous potential of this method, future research could still improve on this topic, particularly in the tasks of object localization and flux prediction.

The first proposed development, which could have a significant impact improving even further the detection tasks, consists in incorporating image defects into the synthetic generation process. Feature such as the inclusion of "hot", "dead" or saturated pixels, or simulating the stitching of images originating from different detectors could provide a significant benefit. These were noted to be regions where the model struggled the most in the real dataset and where it is thought that the largest improvement can be obtained. Techniques to evaluate image similarity, such as those proposed in Chapter 6.3.2, could be employed to guide this development.

For further improvements for both localization of the star centre and flux prediction, two key ideas are proposed. The first consists in a more accurate representation of the stellar PSF, using more elaborate generation techniques and incorporating new functions such as the Moffat or Lorentzian. This could yield a more realistic representation of the pixel region containing the star and allow for more precise measurement of the astrometric properties in these small regions. In conjunction with that, additional customization of the Faster R-CNN architecture could also yield enhanced results. A bespoke architecture specific for the tasks of detection and photometric analysis of stellar objects would allow the network to derive more complex features from the information available. These could later bet used be the prediction head to perform a more precise inference into the positions of the centres as well and the total flux contained in each star region.